Algebraic Reasoning

Professor Arbegla Introduces Variables and Functions

GEMS® Teacher's Guide for Grades 3–5

by
Jaine Kopp
with **Lincoln Bergman**

Skills

Computation (Addition, Subtraction, Multiplication, Division) • Mental Math • Estimation
Creating T-tables • Decoding Patterns • Solving Equations • Decomposing Numbers
Using Variables to Write Algebraic Expressions and Equations • Applying Formulas
Using Models • Problem-Solving • Reflecting • Communicating • Generalizing
Drawing Conclusions • Reasoning and Proving

Concepts

Number Sense • Properties of Numbers • Inverse Operations • Equality • Area
Perimeter • Algebraic Notation (Variables) • Pattern • Function

Themes

Systems and Interactions • Models and Simulations • Patterns of Change

Mathematics Strands

Number • Functions and Algebra • Geometry • Measurement

Nature of Science and Mathematics

Cooperative Effort • Interdisciplinary Connections • Real-Life Applications

Time

Thirteen class sessions of 45–60 minutes; one assessment session of 30–45 minutes

Great Explorations in Math and Science
Lawrence Hall of Science
University of California at Berkeley

Lawrence Hall of Science,
University of California,
Berkeley, CA 94720-5200

Director: Elizabeth K. Stage

**Cover Design, Internal Design, and
Illustrations:** Lisa Klofkorn
Photography: Dan Krauss

Director: Jacqueline Barber
Associate Director: Kimi Hosoume
Associate Director: Lincoln Bergman
Mathematics Curriculum Specialist:
Jaine Kopp
GEMS Network Director:
Carolyn Willard
GEMS Workshop Coordinator:
Laura Tucker
Staff Development Specialists:
Lynn Barakos, Katharine Barrett, Kevin
Beals, Ellen Blinderman, Gigi Dornfest, John
Erickson, Stan Fukunaga, Linda Lipner,
Karen Ostlund
Distribution Coordinator: Karen Milligan
Workshop Administrator: Terry Cort
Trial Test and Materials Manager:
Cheryl Webb
Financial Assistant: Vivian Kinkead
Distribution Representative:
Fred Khorshidi
Shipping Assistant: Justin Holley
Director of Marketing and Promotion:
Steven Dunphy
Principal Editor: Nicole Parizeau
Editor: Florence Stone
Principal Publications Coordinator:
Kay Fairwell
Art Director: Lisa Haderlie Baker

Senior Artists:
Carol Bevilacqua, Lisa Klofkorn
Staff Assistants: Marcelo Alba,
Kamand Keshavarz, Eyad Latif
Contributing Authors: Jacqueline Barber,
Katharine Barrett, Kevin Beals, Lincoln
Bergman, Susan Brady, Beverly Braxton,
Mary Connolly, Kevin Cuff, Linda De
Lucchi, Gigi Dornfest, Jean C. Echols, John
Erickson, David Glaser, Philip Gonsalves, Jan
M. Goodman, Alan Gould, Catherine
Halversen, Kimi Hosoume, Susan Jagoda,
Jaine Kopp, Linda Lipner, Larry Malone,
Rick MacPherson, Stephen Pompea, Nicole
Parizeau, Cary I. Sneider, Craig Strang,
Debra Sutter, Herbert Thier, Jennifer Meux
White, Carolyn Willard

Initial support for the origination and publication of the GEMS series was provided by the A. W. Mellon Foundation and the Carnegie Corporation of New York. Under a grant from the National Science Foundation, GEMS Leaders Workshops have been held across the country. GEMS has also received support from: Employees Community Fund of Boeing California and the Boeing Corporation; the people at Chevron USA; the Crail-Johnson Foundation; the Hewlett Packard Company; the William K. Holt Foundation; Join Hands, the Health and Safety Educational Alliance; the McDonnell-Douglas Foundation and the McDonnell-Douglas Employee's Community Fund; the Microscopy Society of America (MSA); and the Shell Oil Company Foundation. GEMS also gratefully acknowledges the contribution of word-processing equipment from Apple Computer, Inc. This support does not imply responsibility for statements or views expressed in publications of the GEMS program. For further information on GEMS leadership opportunities, or to receive a catalog and the *GEMS Network News,* please contact GEMS. We also welcome letters to the *GEMS Network News.*

Library of Congress Cataloging-in-Publication Data

Kopp, Jaine.
  Algebraic reasoning : Professor Arbegla introduces variables and functions : GEMS teacher's guide for grades 3-5 / by Jaine Kopp with Lincoln Bergman.
      p. cm.
  ISBN 0-924886-70-6 (trade paper)
  1. Algebra--Study and teaching (Elementary) I. Bergman, Lincoln. II. GEMS (Project) III. Title.
  QA159.K67 2003
  372.7--dc21

                                    2003013792

International Standard Book Number: 0-924886-70-6

Printed on recycled paper with soy-based inks.

ACKNOWLEDGMENTS

Professor Arbegla came to life in Ed Martinez's fifth-grade class at Edward M. Downer Elementary school. His students' enthusiastic participation in the *Algebraic Reasoning* activities helped guide the development of this unit, as did Ed's feedback. Our heartfelt thanks and appreciation to Ed and his lively students!

Thanks also goes to Carolyn Willard, who served as my GEMS "buddy" during the first trial test in Ed's class. Carolyn's observations and acumen were invaluable. Special thanks go to my colleague, Terri Belcher, who took on the persona of Professor Arbegla and visited Mr. Martinez's class. Those students will forever view mathematicians through new lenses!

Through my work with Dr. Hung-Hsi Wu, professor of mathematics at UC Berkeley, the mathematical underpinnings of this guide were informed and refined. In addition, Wu's comments and review of the guide, particularly the "Background for the Teacher," were invaluable and appreciated. Thank you, Wu!

Lincoln Bergman has a gift of verse and has crafted a special poem for this guide. Thank you, Lincoln, for that contribution as well as the many other ways that you have enriched this unit. Algebra for All!

The students of Antoineta Franco, Emily Vogler, and Ed Martinez at Downer Elementary grace the photographs in this guide. Thanks to all of these young mathematicians and their dedicated teachers.

Many enthusiastic and supportive teachers across the country served as reviewers for *Algebraic Reasoning* in its trial stages, and we deeply appreciated their valuable comments, classroom ideas, and samples of student work. Your insights and feedback helped us "solve for the variables" and create a cohesive unit. Please see pages 138–140 for a complete list of names.

To the GEMS staff, my thanks and appreciation for your support in its many forms and fashions. I'm proud to be a member of this team, without whom this "gem" would not shine! ∎

Contents

TIME FRAME

Note: The times below are estimates based on classroom testing. The sessions may take less or more time with your class, depending on students' prior knowledge of mathematics, their skills and abilities, your teaching style, and other factors. As will be evident from the step-by-step instructions, and in line with leading mathematics standards, we've placed strong emphasis on classroom discourse/discussion, with students encouraged to articulate their reasoning. These time estimates are in no way meant to limit this important part of effective mathematics teaching and learning.

WHAT YOU NEED FOR THE WHOLE UNIT

The quantities below are based on a class size of 32 students. You may, of course, require different amounts for smaller or larger classes. This list gives you a concise "shopping list" for the entire unit. Please refer to the "What You Need" and "Getting Ready" sections for each individual activity, which contain more specific information about the materials needed for the class and for each team of students.

Nonconsumables

- ❑ 1 copy of Professor Arbegla's **Letter #1: Malfunction in the Function Machine** (page 38) to be signed
- ❑ 1 overhead transparency of *signed* **Letter #1**
- ❑ 1 copy of Professor Arbegla's **Letter #2: Oh, No, Not Again!** (page 39) to be signed
- ❑ 1 overhead transparency of *signed* **Letter #2**
- ❑ 1 copy of Professor Arbegla's **Letter #3: My Multiplication Discovery** (page 84) to be signed
- ❑ 16 sets of base ten blocks★ (wooden, plastic, or paper) in the following approximate denominations (the quantity will vary depending on the magnitude of the numbers used in the activity):
  - __ 35 ones (units) per set for a total of 560
  - __ 15 tens (longs) per set for a total of 240
  - __ 1 hundred (flat) per set for a total of 16
- ❑ 16 bags or containers to hold the block sets
- ❑ 1 overhead transparency of one-centimeter grid paper (page 106, or you can purchase)
- ❑ *(optional)* 8 cups of dry beans (black, pinto, or any small bean)
- ❑ *(optional)* 32 containers for beans (8- or 16-oz. yogurt, sour cream, and/or cottage cheese containers work well)
- ❑ *(optional)* 1 overhead transparency of *signed* **Letter #3**

★You can order sets of base ten blocks (see "Resources" on page 130) or make paper models from the master on page 85.

Consumables

- ❑ 32 algebra (or mathematics) journals
- ❑ 32 *signed* copies of Professor Arbegla's **Letter #2**
- ❑ *(optional)* 32 lists of problem-solving strategies (page 27) to place in journals

General Supplies

- ❑ 5 sheets of butcher paper, each approximately 24 in. x 36 in.
- ❑ several wide-tipped colored markers
- ❑ masking tape
- ❑ an overhead projector
- ❑ overhead pens
- ❑ 33 sheets of construction paper, 12 in. x 12 in.
- ❑ a ruler
- ❑ scissors
- ❑ 1 yardstick or tape measure
- ❑ string
- ❑ approximately 128 sheets of one-centimeter grid paper (page 106 or purchased)
- ❑ *(optional)* 32 thick pieces of cardboard to make a pocket in journal

INTRODUCTION

This GEMS mathematics guide is designed to build a foundation in algebra for students in grades 3–5. And who better to introduce them to algebra than a professor of mathematics! Professor Arbegla (a fictitious colleague of yours) and her "Fabulous Function Machine" help set a compelling context to draw students into algebraic thinking, reasoning, and recording. The function machine launches the unit as students determine a variety of functions the machine performs. Letters from the professor ask your students to help her unravel what appear to be "malfunctions" in her machine. Further on in the unit, they face challenges posed by another mathematics machine that you, their teacher, have invented—a "Morph Machine." Later, they're introduced to Professor LaBarge, who has special scales. Along the way, students' ideas that sum up what they've learned are recorded both on a class chart and in student journals.

As the unit unfolds, students develop algebraic thinking and reasoning skills, and learn laws and properties of our number system. Learning builds in a progression of levels from concrete to connecting to the abstract. This is one of the strengths of the unit—the scaffolding of student learning, so that it builds from activity to activity. There are numerous opportunities to revisit and solidify understandings of the concepts developed. At the start of each activity, the essential content is explicitly stated in a special boxed feature for the teacher.

Algebra Tools and Key Concepts— Activity 1

- Look for patterns
- Make a T-table, or T-chart
- Letters or symbols can represent numbers: $n = 3$ or $\lozenge = 7$
- Variables are letters that represent numbers

Given algebra's strong connection to number and operations, students have many opportunities to practice computational and mental math skills. For example, as they explore the distributive property, students gain a deeper understanding of our base system, practice computation, apply the property to numbers, and then generalize it with variables. Students also use T-tables and variables, express and solve algebraic

Though traditionally algebra instruction began in middle and high school, there has been a major shift in the way mathematics educators see this crucial subject. Now a child's mathematics education is seen as a continuum of learning from pre-kindergarten through high school. In this way, foundations in all the major content areas—including algebra—are established in the early years and built upon throughout a child's mathematics education.

Be sure to provide time for students to think through the problems you present throughout the unit. This helps create persistence in looking for solutions.

expressions and equations, apply inverse operations, identify the commutative properties of addition and multiplication, and investigate area and perimeter through the lens of algebra.

The content in Algebraic Reasoning also addresses many state standards. Check your state guidelines to align the content with your standards. (The California Mathematics Standards, for instance, are particularly well addressed through the activities in this guide.) Though many states have adopted the NCTM Standards, other states have modified them to meet their goals for students.

Many adults' experiences with algebra in high school involved using symbols, following procedures, simplifying expressions, and solving complicated equations—without ever really understanding the **"why"** behind what they were doing. This GEMS unit is the antithesis of that approach; students are actively involved in their own learning every step of the way. The teacher serves as the facilitator of classroom discourse, asking questions and providing content at strategic moments. A class chart is used to document what's learned in each session, creating shared classroom knowledge. In addition, students record the "learnings" in their journals and are often challenged to explain their reasoning. Students work independently, with partners, and as part of the whole class. They continue to think about mathematics outside of class through homework assignments.

Algebraic Reasoning
and
National Content Standards

This guide addresses the National Council of Teachers of Mathematics (NCTM) Principles and Standards of School Mathematics for Algebra and the strand for Numbers and Operations, and reinforces content in the areas of Number Sense, Geometry, and Measurement. Principles and Standards emphasizes the need to teach elements of all content areas of mathematics from pre-kindergarten through high school. As stated, algebra is no exception:

> By viewing algebra as a strand in the curriculum from pre-kindergarten on, teachers can help students build a solid foundation of understanding and experience as a preparation for more-sophisticated work in algebra in the middle grades and high school. For example, systematic experience with patterns can build up to an understanding of the idea of function, and experience with numbers and their properties lays a foundation for later work with symbols and algebraic expressions....

Specific to the grade range for which this GEMS guide is designed, grades 3–5, Principles and Standards provides additional guidance as follows:

In grades 3–5, students can investigate properties such as commutativity, associativity, and distributivity of multiplication over addition.... An area model can help students see that two factors in either order have equal products. At this grade band, the idea and usefulness of a variable (represented by a box, letter, or symbol) should also be emerging and developing more fully. As students explore patterns and note relationships, they should be encouraged to represent their thinking. As students become more experienced in investigating, articulating, and justifying generalizations, they can begin to use variable notation and equations to represent their thinking. Teachers will need to model how to represent thinking in the form of equations. In this way, they can help students connect the ways they are describing their findings to mathematical notation. Students should also understand the use of a variable as a placeholder in an expression or equation. In grades 3–5, algebraic ideas should emerge and be investigated as students—

- extend numerical and geometric patterns

- describe patterns verbally and represent them mathematically in words, symbols, tables, and graphs

- look for and apply relationships between varying quantities to make predictions

- express mathematical relationships using equations

- identify commutative, associative, and distributive properties and use them with whole numbers

- represent the idea of a variable as an unknown quantity using a letter or symbol

- investigate how a change in one variable relates to a change in a second variable, such as Area (of a rectangle) = $l \bullet w$

- make and explain generalizations that seem to always work in particular situations

- model problem situations with objects and use representations such as graphs, tables, and equations to draw conclusions

- explore number properties

- use invented notation, standard symbols, and variables to express a pattern, generalization, or situation

While no single series of activities can take on all algebraic underpinnings, Algebraic Reasoning focuses in depth on several key elements. The "Background for the Teacher" on page 107 provides more detail on specific concepts exemplified in this unit and aligned with Principles and Standards, relating to both the Algebra Strand and the Numbers and Operations Strand for grades 3–5.

Algebra: An Ally for All Students' Success

Algebra is a truly fascinating branch of mathematics and one of the most amazing inventions of the human mind! When introduced in a positive way, and in an interesting context, it can draw out the creative, problem-solving, and puzzle-figuring-out abilities of all students. There's a special feeling of challenge and triumph that comes in solving for an unknown.

Algebra's real-world connections abound, and students' confidence in mathematics, success in higher education, and future career opportunities can all be deeply influenced by their encounters with algebra. Unfortunately, for a variety of reasons, algebra has attained a cultural reputation as complicated, difficult, and generally for "smart" students. This GEMS unit can be one small but powerful part of a larger and much-needed social and educational process to improve the teaching of and access to algebra. It does so by sparking interest in algebraic reasoning, making it understandable and encouraging students to grapple with it, gaining positive experiences in grades 3–5. (The GEMS unit *Early Adventures in Algebra*, for students in grades 1 and 2, begins the process even with these young children.)

Taken together, the activities in this unit will help your students build a solid foundation for more advanced algebraic reasoning in later grades. Though students are likely to progress at different rates and in different ways (which is to be expected), *all* students should be able to grasp the concepts in this guide, and to do so without the anxiety often associated with algebra.

Activity-by-Activity Overview

The unit opens with an invitation to students to express what they've heard about algebra and to ask any questions they may have about it. Each student receives a special "Algebra Journal" to document her work and learning throughout the unit.

In **Activity 1: The Fabulous Function Machine,** students are introduced to Professor Arbegla and her amazing machine. They suggest numbers (and later geometric shapes) to go *into* the machine, and then analyze what comes *out*. Students learn to look for **patterns** as a key strategy to decode the "secret rule" the machine is following for each

---

Algebra got its name more than a thousand years ago, from the title of a mathematical work by scholar Mohammed ibn-Musa al-Khwarizmi. See "Background for the Teacher" (page 107) for more on the history of "algebra."

Algebra is one of the primary "gatekeepers" in tracking students along academic or non-academic pathways, affecting their entire lives. See "Background for the Teacher" (page 107) for more on this important strategic reality.

new function. Students are introduced to the use of *T-tables* to organize the data, as well as the use of *variables* to write algebraic expressions for the computational operations the machine is performing. Students begin to build skills, strategies, and tools that will help them do algebra. These are recorded on an **Algebra Tool Kit** class chart as well as in the students' journals.

In **Activity 2: Malfunctions in the Function Machine,** Professor Arbegla's invention appears to be acting up. She sends a letter to your class, describing the problem (the same number that goes in, comes out) and asking your students for assistance. This "malfunction" introduces the class to the *identity element for addition and/or multiplication.* Students may also discover other computational operations, in which a number goes into the machine and comes out the same. After the class has solved this problem, a second letter arrives from the professor, describing a new "malfunction." This time, no matter what number goes into the machine, the number seven comes out. As homework, students are challenged to figure out what the machine is doing.

In **Activity 3: The Morph Machine,** students are introduced to a new, two-step machine that you yourself have created. When students put numbers into the Morph Machine, the numbers go into the "trans-former chamber" and a series of operations are applied on them. Then you put the resulting "morphed" numbers into the machine's "restorer chamber" and—"magically"—are able to determine the original num-bers that went in. The magic is the use of *inverse operations,* or simplify-ing an algebraic expression. In the second session, you share the secret of your machine with your students.

In **Activity 4: Professor LaBarge's Scales,** the use of a scale provides the context to *solve equations for one or more variables.* In this model, the scale represents an equation and the weights represent numbers and/or variables. In the first session, students are introduced to the scale itself, given some of the weights on the scale, and asked to find the value(s) of the unknown weights. They use variables to express these equations algebraically. In the second session they're given more equations to balance—some with a single solution and others with an infinite set of solutions. As they solve these problems, students are introduced to the *commutative property of addition.*

In **Activity 5: The Distributive Property,** another letter arrives from Professor Arbegla. This time there's no problem to solve! Instead, the professor shares her "great multiplication discovery," known in mathematics as the ***distributive property,*** or law of operation. In Session 1, students are introduced to the distributive property over addition, and have opportunities to apply it to problems. At the end of the first session, students use algebra to generalize this property. In Session 2, students discover that there's also a distributive property over subtraction, and learn to write its algebraic expression. Finally, they apply what they've learned about both distributive properties as they solve a variety of problems. The properties are recorded on the **Algebra Tool Kit** chart and in students' journals.

In **Activity 6: Algebra in Action,** students explore the ***area and perimeter functions*** for rectangles. In the first session, they're introduced to the ***commutative property of multiplication*** as they investigate a room with an area of 36 square feet and see the impact of length and width on the room's size. In the second session, students create a standard unit of measure, a square foot. Using that unit, the class concretely measures and maps to scale the length and width of an enclosure with an area of 36 square feet, to see its actual size. Using string, they measure the perimeter of the rectangle, and distinguish between the area and perimeter measurements. In the final session, students solve area problems with variables to ***find an unknown in the equation***—the length, width, or area. As students explore problems involving both area and perimeter, they have an opportunity to see how distinct these two functions are.

Special Features

Algebra Tool Kit

Over the course of the unit, students are encouraged to explain their thinking and write in **algebra journals.** To support their learning of key concepts in algebra and number, an "Algebra Tool Kit" is created and developed. The feature isn't an actual kit of materials, but a classroom chart that summarizes tools the class accumulates as the unit progresses. This chart is an important resource to keep content knowledge accessible to all students. It grows from listing tools, such as a T-table to help determine a function, to including key concepts and properties, such as the commutative property of addition and multiplication. Everything recorded on the chart flows from the class activities and discussions. Students also create their own tool kits in their journals, so they can have direct personal reference to these helpful tools.

Sample Dialogues

Like all GEMS guides, *Algebraic Reasoning* provides clear step-by-step presentation instructions. In this unit, we've added something more to help you bring out the best in your students and to emphasize how important classroom discourse is in mathematics learning. We've included some sample teacher-student dialogues, **based on real classroom interactions,** and placed them near the concepts to which they relate. These sample dialogues also provide examples and mini-case studies of the various discussions and back-and-forth questioning likely to occur as you present the unit, and are intended to help you encourage students to express their thinking out loud and explain their reasoning whenever possible. They offer insight into some good ways to facilitate such classroom discourse.

Alge-Branch Points for Teachers

This guide is written for grades 3 through 5. At certain points in presentation, you'll need to adapt an activity depending on your students' abilities. We've identified these junctures as "Alge-Branch Points" and set them off from the main text so you can see them coming and plan your strategy.

Additional Assessment Activity

What Comes Next? (page 128) is designed as a concluding assessment activity to help you gauge how students have understood the content of the unit, and how well they apply what they've learned. Student responses can serve to guide the additional work you do with them on concepts and skills developed in this unit.

Algebra Vocabulary for the Teacher

"Background for the Teacher," page 107, includes glossaries and definitions of algebraic terms, operations of numbers, and properties. This can provide a good refresher before you begin the unit, and/or serve as a reference throughout. ■

Overview

This activity's two sessions introduce students to algebraic reasoning and recording through a familiar teaching tool—the function machine. A twist is added to draw students in—the introduction of Professor Arbegla, the inventor of the Fabulous Function Machine. Students are challenged to determine what the machine is doing to the numbers or shapes that enter it, depending on the function, and to write its algebraic expression. The class begins to put what they've learned into an **Algebra Tool Kit**—a class chart and the "tools," or key concepts, that will fill it over the course of the unit: mathematical definitions, procedures, problem-solving strategies, properties and algorithms.

In Session 1, students share what they know or have heard about algebra; their ideas and questions are recorded on charts that remain posted throughout the unit for reference as questions are answered or ideas are confirmed. Students are then introduced to a "colleague" of yours; Professor Z. Arbegla, a professor of mathematics. Professor Arbegla has shared her Fabulous Function Machine to try out with your class. Students volunteer numbers to put into the machine, and you "operate" the machine and, using one function at a time, give the output. The goal is to get the students familiar with the machine and create excitement about determining which function is at work—its "secret rule."

In Session 2, you continue to provide functions for the class to solve. As students discover each "secret rule" for the machine, they determine what the machine is doing. To provide scaffolding for the functions presented in Activity 2, it's suggested that one function you present be multiplying by zero.

The box on page 10 shows some of the tools and concepts you can record on your class chart with your students, depending on the content you cover with them. This makes the learned content concrete and available to everyone. As much as possible, the contents of the chart should be generated by your students. *These advance glimpses are just for the teacher;* they show some of the tools students will have acquired by the end of the activity. It's important not to add properties or concepts to the **Algebra Tool Kit** chart until students have been introduced to them.

Functions are described in "Background for the Teacher," page 107.

In Algebraic Reasoning we've used a dot (•) instead of "x" to indicate the multiplication operation in algebraic expressions. This is a good habit to form with your upper-elementary students, who'll encounter this mathematical convention at higher grades and in textbooks. If your students are more comfortable using an "x" to denote multiplication, feel free to use that notation.

When you record letters as variables, use lowercase letters; it's the convention in later grades. When "x" is used as a variable, it's italicized.

Session 1:
Professor Arbegla's Function Machine

■ What You Need

For the class:
- ❑ 3 sheets of butcher paper, each approximately 24 in. x 36 in.
- ❑ several wide-tipped colored markers
- ❑ masking tape

For each student:
- ❑ 1 algebra (or mathematics) journal
- ❑ *(optional)* 1 thick piece of cardboard to make a pocket in journal
- ❑ *(optional)* list of problem-solving strategies (page 27) to place in journal

■ Getting Ready

1. On two sheets of butcher paper, write headings as follows:

- chart 1: **Our Ideas about Algebra**

- chart 2: **Our Questions about Algebra**

2. On the third sheet of butcher paper, draw an illustration of the Fabulous Function Machine using this graphic as a guide. This is your **Fabulous Function Machine** chart for the class.

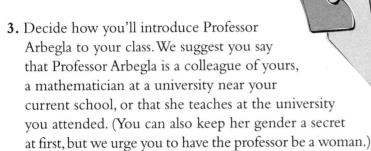

3. Decide how you'll introduce Professor Arbegla to your class. We suggest you say that Professor Arbegla is a colleague of yours, a mathematician at a university near your current school, or that she teaches at the university you attended. (You can also keep her gender a secret at first, but we urge you to have the professor be a woman.)

4. Make or purchase an algebra journal for each student. **They'll use the journal throughout the unit.** If you choose, tape the thick cardboard on the inside back cover of each journal to act as a pocket to hold loose homework.

5. You may wish to copy and cut apart the list of problem-solving strategies on page 27, for students to place in their journals and refer to over the course of the unit. We suggest distributing them at the end of Session 2.

6. Before you present the session, gather markers, masking tape, strategy lists, and the three charts.

We've found that a special journal for this topic adds to the excitement and enthusiasm—it gives algebra status! However, if your students already have a mathematics journal, feel free to have them use that one.

 ■ **Introducing Algebra**

1. Tell students you have a special journal for them for a new unit in mathematics. Distribute the journals and have students write their names on them.

2. Write "ALGEBRA" on the board. Ask for a show of hands from anyone who's ever heard of algebra.

3. Tell your students you know they bring a lot of knowledge with them into the classroom. You want to find out what they know or have heard about algebra before you begin the unit.

4. On the first page of their journals, have them write what they know about algebra and list any questions they may have about it. Emphasize that it's fine to include things they've heard about, *whether or not* they're sure those statements are correct. It's also fine to express part of an idea, and they shouldn't worry about getting a "right answer."

5. On the board, write some questions to assist them, such as:

- What do you know about algebra?

- What have you heard about algebra?

- How do you think algebra is used?

- What questions do you have about algebra?

6. Create a completely quiet classroom environment so students can think and write independently, and have them begin.

7. While they're writing, post the charts **Our Ideas about Algebra** and **Our Questions about Algebra.** Have the **Fabulous Function Machine** chart ready to post.

8. Circulate as students are writing. Allow about 10 minutes of writing time.

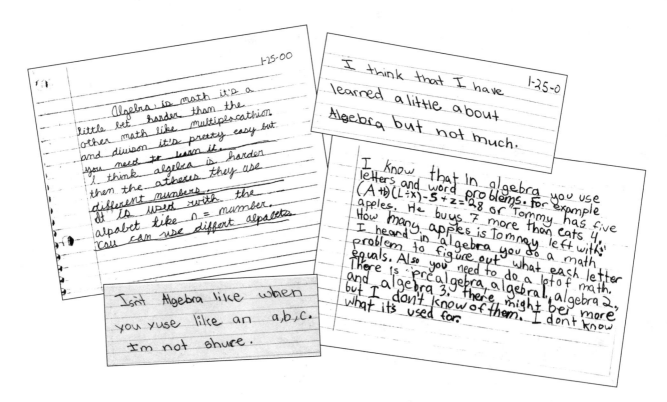

9. Focus the class on the two charts. Tell the class you want to record their ideas and questions on the charts. Again, assure students that they can say things they've just heard about; they don't have to be 100-percent sure of their ideas.

10. Begin calling on students and recording their thoughts on the **Ideas** chart. Very often a student will open with, "I've heard that algebra is hard." Write that on the chart—"Algebra is hard." Ask if others have also heard that algebra is hard. Chances are that many have!

It's difficult to know exactly how students will respond. See pages 14–15 for some samples of student responses at different grade levels, and ways to follow up on some of their ideas.

11. Next, move on to their questions about algebra. Record them on the **Questions** chart. Some examples include: "Does it have to have numbers, or could it be all letters?" "How do you use it?" "How does it work?" Some classes will have many more.

12. Tell students that over the course of the unit, as they learn more about algebra, you—or they—may refer to the charts to see if any of the ideas need revision or if they've learned the answers to some of the questions.

Sample Student Responses:
"Our Ideas about Algebra"

"Instead of numbers, it's got letters."

Teacher response:
- Ask student for an example.
- Student says "$x + y$."
- Write "$x + y$" on the chart.

"It uses variables."

Teacher response:
- Ask student what she means by "variable."
- Student says "x is a variable."
- Ask her for an example.
- Students says "5 times x."
- Record "$5 \cdot x$" on the chart and check that this is what the student meant.

    or

- Have the student come up and record her own idea.

"When you have the answer, you put the letter equals the answer."

Teacher response:

- Ask that student for a letter.

- Student says "*n*."

- Write "*n* = _____" on the chart.

- Check that this is what he meant, or have him come up and record his idea.

"It's kinda like math."

Teacher response:

- Write "Algebra is like math" on the chart.

"It's like a number sentence."

Teacher response:

- Ask, "What is a number sentence?" Give students time to discuss and come up with examples.

- Have table groups or partners discuss this idea.

Sometimes your students will know quite a bit about algebra. If that is the case, the discussion may be more like these grade-level specific examples:

In a third-grade class, you may be given an answer like this: $45 + \Diamond = 50$

Teacher response:

- Ask students if they know what value "$\Diamond$" needs to equal to make the equation true.

- Have them talk with a partner about their answer.

- Substitute the number for "$\Diamond$" and solve.

In a fourth-grade class, you may be given $25 - x = 6$

Teacher response:

- Have them discuss this number sentence with partners.

- Have a student give an answer and explain how she got it.

- Substitute the number and solve.

In a fifth-grade class, one student gave the following answer:
$y \cdot 1 = y$

Teacher response:
- Ask other students what that could mean.

- Have them explain, then record it on the chart.

Note: It's OK if not everyone understands at this moment.

Another fifth-grade student gave the following example: $(z \cdot 2) + 4 - 5 = 15$

Teacher response:
- Provide time for students to solve this problem and propose answers.

- Substitute the values they give for "z" and solve the equation.

NOTE: Once students understand this idea, they're likely to be able and eager to write other number sentences. After a few number sentences, give the students a chance to record a number sentence in their journals and then guide them to think about their other ideas about algebra.

■ The Fabulous Function Machine

1. Introduce Professor Arbegla in the way you've chosen. Say that this professor is always doing interesting projects.

2. Post the **Fabulous Function Machine** chart near a chalkboard and say this is one of Professor Arbegla's inventions—a Fabulous Function Machine. Pointing to the arrows going in and out of the machine, say that when things go in, something happens to them before they come out. You're going to demonstrate it for them.

3. On the board near the picture of the machine, draw a T-table to record the numbers students will suggest and the output of each.

IN	OUT

4. Ask students if they know that this is a T-table, and what it's for. If needed, clarify that it's a tool to organize numbers—in this case, the numbers that go into and come out of the Function Machine.

5. Ask for a number from 1 through 50 and record it under the "IN" part of the T-table. Point to the **Fabulous Function Machine** chart, indicating that the number is going into the machine to be pro-

IN	OUT
7	1
25	1
10	0
15	1
4	0
1	1
12	0

It's likely that students will get very excited as numbers come out. They'll probably want to guess the function immediately. Try to steer them in the direction of providing HINTS to their classmates so the other students can "get it" as well.

cessed. Make some physical signs as you process the number—close your eyes, rock back and forth, make a low machine sound, blink your eyes...be creative!

6. If the number you were given was even, write "0" as the output. If the number was odd, write "1" as output.

7. Continue taking numbers, pointing to the chart, processing the numbers, and recording the outputs. Tell the students their challenge is to *silently* figure out the "secret rule" for what the machine is doing.

8. When you feel that several students have figured out your "secret rule," ask someone who thinks she knows the rule to give a **hint**— not tell the rule—so her classmates can also figure it out. Put a number in and let students tell the person sitting next to them what they think will come out.

9. When about half the class seems to know the rule, ask someone to explain what the machine is doing. Check that the rest of the class agrees.

10. Give a number and ask students to whisper to their neighbors what number they think will come out. Ask a student for the answer and record the output. Take one final number to go in, and again let students tell the output.

Alge-Branch Point for Teachers

At this point you'll need to adapt the activity depending on your students' abilities. The next function that the machine performs should be a one-step addition function. With **third-graders,** a "+3" function would be appropriate, while **fifth-graders** may be able to handle a "+11" function.

■ Introducing the One-Step Addition Function

1. Tell students they're now ready for a bigger challenge, so you're going to *reprogram* the machine. Draw a new T-table on the board and have students draw one in their journals.

2. Ask for a number from 1 through 25. Record it on the T-table and have students record it in their journals. "Process" the number and record the output appropriate for your function.

3. Continue asking students for numbers and processing them in the machine. As you record the input and output numbers on the T-table on the board, have students record them in their journals.

4. When about half of the students have figured out the function, record a number only under the IN part of the table. Have students record it and write the predicted output in their journals. After they've written their ideas, have them share outputs with the person sitting next to them.

5. Ask the class for the output with an explanation of what the machine is doing. Encourage students to come to the board and use the T-table to explain how they figured out the machine's "secret rule."

6. Post the **Fabulous Function Machine** chart in a place where it can remain throughout the unit, for reference. Leave the **Ideas** and **Questions** charts posted for future reference as well. Collect the students' algebra journals.

Reviewing your students' journals after each activity helps you monitor their understanding of the concepts you've covered. This, in turn, will help you choose and pace the functions you'll introduce as the unit progresses.

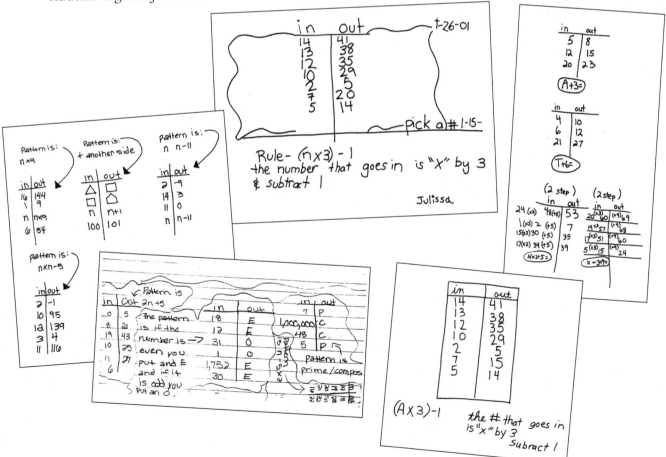

Session 2: Decoding Functions

■ What You Need

For the class:
- ❏ the **Fabulous Function Machine** chart from Session 1
- ❏ 1 sheet of butcher paper approximately 24 in. x 36 in.
- ❏ several wide-tipped colored markers
- ❏ masking tape

For each student:
- ❏ algebra journal from Session 1

■ Getting Ready

As the unit progresses, it's likely you'll need to add sheets of butcher paper to record the accumulating mathematical tools used in the unit.

1. On the butcher paper, prepare a chart for the **Algebra Tool Kit** using the graphic below as a guide.

2. Decide which functions you'll use with your class after the "geometric shapes" function. See the Alge-Branch and More Functions to Figure Out on page 23 to select functions that meet the needs and levels of your students.

3. Before you present the session, gather markers, masking tape, journals, and the **Algebra Tool Kit** chart.

 ### ■ Geometric Shapes and the Function Machine

1. Pass out the journals to the students and tell them you have some new things for them to put into the Function Machine. Start by asking what "geometry" is. Provide a few hints if needed.

2. After a bit of class discussion, agree that geometry is the study of two- and three-dimensional shapes. Tell them that today a special type of shape is going into the Function Machine—a **polygon.** Ask what a polygon is. Collaborate to come up with a clear descriptive definition: *A polygon is a closed two-dimensional shape with three or more sides.*

3. Explain that when a polygon goes into the Fabulous Function Machine, it comes out transformed. Every time a polygon enters the machine, it comes out as a polygon on which **the number of sides has increased by one.** A triangle will come out a quadrilateral; a quadrilateral will come out a pentagon, and so on.

4. Ask a student to name a polygon. Make a T-table and under the IN part draw the polygon suggested by the student. (*Note:* There are many polygons students could suggest as input to the machine— you'll need to adjust the *output* according to the function. See page 110 for examples of polygons students might come up with.)

The following dialogue suggests ways to embed geometric learning into class discourse.

*The definition of a polygon could go on the **Algebra Tool Kit** chart.*

Since there are many quadrilaterals, you can choose any one to come out—a rectangle, square, parallelogram, rhombus, etc. Students can suggest quadrilaterals to go into the machine, and you can vary the pentagons (orientation, length of sides, etc.) that come out. The function stays constant; the number of sides of the polygon that goes in increases by one.

Sample Dialogue:
Geometric Shape Function

Teacher (TR): Who has a polygon they want to go in the machine?
Charles: A square.

TR draws the square in the T-table under IN, makes the machine operate, and draws a pentagon as output.

TR: Who can give me another shape?
Maria: A diamond.
TR: Do you know what a diamond looks like? How many sides does it have?
All: 4!

TR draws a diamond on the T-table.

TR: Many people call this shape a diamond, but it also has another name—a geometric name. Does anyone know what it is?
Adriana: A rhombus.
TR: Why is it a rhombus? Tell the person sitting next to you.
TR: Can someone give us a definition of a rhombus?
Adriana: It's a parallelogram and it has 4 equal sides.
TR: Is there anything else anyone wants to add?
Sydney: I think the angles are the same.

TR: Which angles are the same? Please come to the board and show us.

Sydney goes to the board and draws a rhombus. She points to the opposite angles and says these two are the same and these two are the same.

TR: What does Sydney mean by "the same"?
Elvis: The angles are equal.
TR: OK, so a rhombus is a special type of parallelogram whose sides are all equal length and whose opposite angles are equal.

TR draws a pentagon as output for the rhombus (diamond).

TR: Who has another polygon to go into the machine?
Tyrell: A triangle.

TR draws a triangle in the T-table under IN and draws a square under OUT.

Joe: Put a rectangle in!

TR draws a rectangle under IN.

TR: What do you think will come out? Talk to the person next to you [or your table mates].

Provides brief time to talk.

TR: Who has an idea of what will come out of the machine?
Sara: A hexagon.
TR: How many agree that a hexagon comes out?

Show of hands.

TR: Does anyone have a different idea about what will happen?
Spenser: I think it'll be an octagon when it comes out.
TR: Who agrees/disagrees with octagon? Any other ideas?

No hands up.
TR draws a pentagon under OUT side of T-table.

TR: Does anyone recognize this shape? Does it look like anything you've seen before?
CJ: It's like home plate on a baseball field!
Celeste: It looks like a house to me.
TR: Does anyone remember its geometric name?

No responses.

TR: Talk to a neighbor.

Provides a moment to discuss.

TR: Any ideas now?

No volunteers.

TR: Has anyone heard of a pentagon?

All: Oh, yeah! Now I remember…
TR: Let's look at our T-table. Does anyone have an idea about what the machine is doing?

No ideas yet.

TR: Give me another polygon.
Ben: A circle.
TR: Is that a polygon? Why or why not?
Lamar: Circles don't have sides, so they can't be polygons.
Ben: Oh, yeah, that's right.
TR: Another shape to go in?
L'Toya: Pentagon.

TR draws a pentagon under IN.

TR: Predict what will come out—whisper your idea to your neighbor.

TR draws a hexagon under OUT side of the T-table.

All: I knew that! Yeah!
TR: Here's your next step. Tell a neighbor [or people at your table] what you think the rule is. See if you agree with one another or have different answers.

TR circulates as students discuss.

TR: Is any pair [or group] ready and willing to share their idea about the rule?
Chenille: It gets one more side.
TR: Explain a little more. What has one more side?
Chenille: Like the pentagon goes in and it's got 5 sides. Then when it comes out, it's got 6 sides. That's one more side.
TR: How many people agree with Chenille's idea?

Show of hands.

TR: Did anyone have a different rule or idea of what's happening?

No hands up.

TR: Let's take a closer look at Chenille's idea. One thing that may help is a new T-table.

TR draws a new table.

TR: Let's start with the polygon with the fewest sides. What is it?
All: Triangle.
TR: We know a triangle has 3 sides. Let's write 3 on the new T-table under IN. What came out?
All: A square.
TR: How many sides does a square have?
All: 4!

TR records 4 as output on the new T-table.

IN	OUT
3	4

TR: What polygon would follow the triangle?

All: A square.

TR writes 4 under IN.

TR: Why did I write 4 under the IN side of the T-table?
Trevor: 'Cause a square has 4 sides.
TR: What should I record under OUT?
All: 5.

TR records 5 under OUT.

IN	OUT
3	4
4	5

Eryka: A rectangle goes in and a pentagon comes out for that too.
Miles: And for a rhombus too...a pentagon come out!
TR: So, when any polygon with four sides went in, what shape came out?
All: Pentagons!
TR: How many sides do pentagons have?
All: 5!
TR: Are we seeing a pattern here?

Using the T-table with numbers to record the pattern of growth of the geometric shapes provides an opportunity to begin using algebraic notation.

In one class, a student asked that a six-pointed star be put into the machine, thinking a seven-pointed star would surely emerge. She was surprised when the output was a 13-sided figure. After students determined what the machine was doing, the class looked at the six-pointed star again and counted its 12 sides. This gave them a chance to review its geometric name—a **dodecagon.**

5. After several polygons have gone into the machine and been transformed, have students look at the T-table. Ask them to describe what's happening. Make a second T-table showing the number of sides of the polygons.

IN	OUT
3	4
4	5
4	5
4	5
5	6

6. Have students explain the pattern in several ways. Students can generally see that for every shape that goes in, the output is one number of sides greater. In addition, they see that for *every* shape with 4 sides that goes in, a shape with 5 sides always comes out!

7. Ask what would happen if you put in a shape with 21 sides. They're likely to rapidly tell you a shape with 22 sides will come out. Give another example, such as a shape with 99 sides. Out would come a shape with 100 sides.

8. Since students know that the rule is add "1" to any number that goes in, show them there's a way to record that algebraically. **Reiterate that no matter what number goes in, it will come out larger by one.**

9. Write "n" under IN on the T-table. Say that "n" represents **any number.** Show them how to record what happens to the number in the machine using algebra: write "$n + 1$" under OUT on the table.

10. Tell students that "$n + 1$" is an algebraic expression for what the function machine is doing. Have students record this T-table in their journals.

11. Offer a number to put IN and ask what will come OUT. This helps develop their understanding of algebraic expressions—in particular, how to substitute a value for a variable.

IN	OUT
3	4
4	5
4	5
4	5
5	6
⋮	⋮
21	22
⋮	⋮
99	100
⋮	⋮
n	$n + 1$

Alge-Branch Point for Teachers

At this point you'll need to adapt the activity depending on your students' abilities. The following guidelines can help you decide how to continue with "Decoding More Functions," on page 24.

■ More Functions to Figure Out

1. Tell students you're going to program a new function into the machine for them to figure out. Depending on their understanding of the first part of this session, and their skills and abilities, select a function that will be challenging but not frustrating. For example, by grade level:

   a. For **third-grade students:** Have the Function Machine simply **add a constant number** to the number that goes in; for example, "plus 4." Also, limit the numbers that go in by defining the domain—say, numbers from 1 through 25. This will help students look for a pattern.

   b. For **fourth- and fifth-grade students:** Have the machine **do a two-step function.** Say that when the number goes in, the function machine does *two different things* to the number before it comes out. Limit the domain to, for example, numbers from 1 through 25. A function that works well for the first try at a two-step function is "double the number and add one" or "$2n + 1$."

   c. For **experienced fifth-grade students:** Have the machine do a more challenging two-step function, such as "$3n - 1$."

A note of caution if you use a function that involves subtraction. Be sure it doesn't produce a negative number for students who have not had experience with that concept. For example, in the case of "x minus 3," if a 1 went in, a "–2" would come out.

2. We recommend you include the function $n \cdot 0 = 0$ for all grade levels. This function will help scaffold the students' decoding of more challenging functions in Activity 2, when the machine malfunctions.

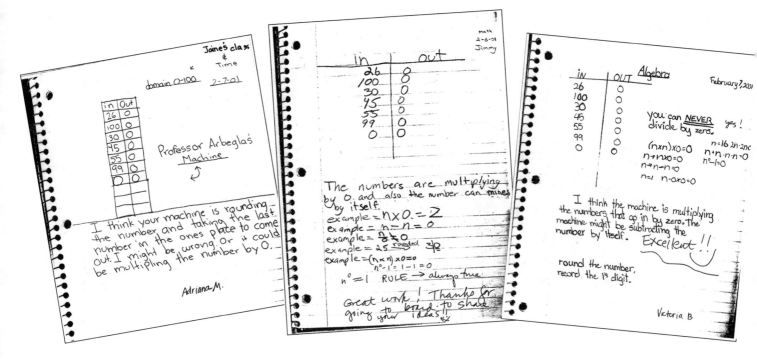

3. Here are some additional functions you might want to use:

One-Step	Two-Step
$2n$	$2n + 7$
$5n$	$5n + 2$
$n + 7$	$7n - 2$

Use the "Sample Dialogue: Geometric Shape Function" (page 19) as a guide to support active student participation in decoding the function.

■ Decoding More Functions

1. Continue choosing functions to "process" through the machine. Regardless of which functions you choose, use a class T-table to record the student-generated inputs and the numbers that come out. Have students record the T-tables in their journals.

2. It may be helpful to make a second T-table next to the original one, to reorganize the numbers that went into the machine from lowest to highest. This helps students identify patterns.

3. When most students seem to know the secret rule, have them use their journals to write what they think the function machine is doing or describe any patterns they see in the T-table.

IN	OUT
5	14
22	65
1	2
10	29
25	74
9	26

(completed in order of numbers given by students)

IN	OUT
1	2
5	14
9	26
10	29
22	65
25	74

(reorganized from lowest to highest)

4. Give them a minute to talk with a partner or small group about how the machine is working. Have them come to agreement about the function.

5. Agree on a letter, such as "*n*," to represent any number that goes into the machine. Ask if anyone knows how to write what the machine is doing to the number in an algebraic expression.

6. If students have trouble, walk them through the process of writing the function. Use words as well as the algebraic symbols. For example, "$2n + 1$" translates into words as "two times any number that goes in plus one more." Have them record in words what the function is doing.

7. Continue to pose functions for them to decipher. As each function is defined, write its algebraic expression and describe it in words.

■ Tools for the Algebra Tool Kit

1. Tell students that people use tools all the time to make their jobs easier. Give (or ask for) examples of a few professions and the tools needed in each (a carpenter, a chef, a medical doctor, a secretary, a teacher, etc.). Say that mathematicians, too, use special tools to help them solve problems, write in mathematical language, and investigate new areas of math.

2. Post the **Algebra Tool Kit** chart. Explain that in the case of mathematicians, the tools aren't hammers or measuring cups or stethoscopes. Math tools include ***mathematical definitions, procedures, problem-solving strategies, properties, and algorithms.*** Ask the students to remember what tools they used to figure out and record the "secret rule" in the Function Machine. Suggested tools might include:

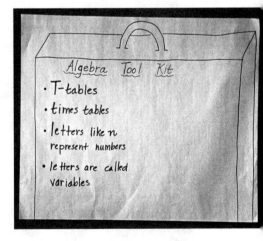

- Using a T-table (or T-chart)

- Organizing the data in the table

- Looking for patterns

- Working with a partner

- Using letters to represent numbers

- Writing algebraic expressions

- Any number times zero is always zero: $n \cdot 0 = 0$

3. If students have trouble coming up with tools, ask leading questions related to the activities they've done.

4. Have them turn to a page in their journal to draw their own Algebra Tool Kit, recording the tools you added to the posted chart.

5. If you choose to use them, distribute the list of problem-solving strategies.

6. Collect the students' journals.

■ Going Further

Have students devise functions for the machine and share them with the class.

Problem-Solving Stategies

- Guess & Check
- Use Manipulatives
- Draw a Picture
- Look for Patterns
- Make an Organized List
- Round Numbers
- Make a T-table or T-chart
- Simplify the Problem
- Work Backwards
- Make a Graph
- Work with Others

Problem-Solving Stategies

- Guess & Check
- Use Manipulatives
- Draw a Picture
- Look for Patterns
- Make an Organized List
- Round Numbers
- Make a T-table or T-chart
- Simplify the Problem
- Work Backwards
- Make a Graph
- Work with Others

Problem-Solving Stategies

- Guess & Check
- Use Manipulatives
- Draw a Picture
- Look for Patterns
- Make an Organized List
- Round Numbers
- Make a T-table or T-chart
- Simplify the Problem
- Work Backwards
- Make a Graph
- Work with Others

Problem-Solving Stategies

- Guess & Check
- Use Manipulatives
- Draw a Picture
- Look for Patterns
- Make an Organized List
- Round Numbers
- Make a T-table or T-chart
- Simplify the Problem
- Work Backwards
- Make a Graph
- Work with Others

Overview

This activity's two sessions kick off with letters from Professor Arbegla, who fears her Fabulous Function Machine may be malfunctioning and asks your class for help. In the first instance, every time the professor puts a number in, that same number comes out. Is the machine broken? What could it be doing to the number to make it come out the same? In the second apparent malfunction, the professor says that no matter what number goes into the machine, the number "7" comes out. How could this be?

Students address the first "malfunction" by theorizing what could be happening to the machine, first with a partner and then in a letter back to Professor Arbegla. Since there's more than one explanation, this problem provides the context to express algebraically several operations the machine could be performing. After they've written the letter in their journals, you guide a class discussion with probing and clarifying questions to build on students' thinking. This problem guides students to concretize and algebraically express and name the ***identity elements for addition*** (zero) ***and multiplication*** (one). By posing expressions such as these, students demonstrate their understanding of numbers and operations.

For the second apparent malfunction in the machine—no matter what number goes in, the number "7" comes out—students are given a copy of the professor's letter to take home to work on independently, using the "tool kits" in their journals to help them solve the problem. When they bring their results back to class, they share their findings first in small groups, then in a class discussion.

In these two sessions, students reinforce the concept of a variable, discern functions, create and write algebraic expressions, concretize the identity elements for addition and multiplication using symbolic notation, and extend their understanding of the number zero. In the course of working through these problems with the class, you'll add new "tools" to the **Algebra Tool Kit,** and students will add to their journals.

Variables, equations, and the identity elements for addition and multiplication are described in "Background for the Teacher."

**Algebra Tools and Key Concepts—
Activity 2**

- Identity Element for Addition: $n + 0 = n$ also written as
  $0 + n = n$ or $n = n + 0$

- Identity Element for Multiplication: $n \cdot 1 = n$ also written
  as $1 \cdot n = n$ or $n = n \cdot 1$

- Other expressions that result in "$n = n$":
  $$n - 0 = n$$
  $$\frac{n}{1} = n$$

Session 1: What Goes In Comes Out (Letter #1)

■ What You Need

For the class:
- ❑ the **Fabulous Function Machine** chart from Activity 1
- ❑ the **Algebra Tool Kit** chart from Activity 1
- ❑ 1 copy of Professor Arbegla's **Letter #1: Malfunction in the
  Function Machine** (page 38) to be signed
- ❑ 1 overhead transparency of *signed* **Letter #1**
- ❑ an overhead projector
- ❑ several wide-tipped colored markers

For each student:
- ❑ algebra journal from Activity 1

■ Getting Ready

1. Copy and trim Professor Arbegla's **Letter #1: Malfunction in the
   Function Machine** (page 38) and have a colleague sign it. (You're
   welcome to adapt the letter if you wish.)

2. Make an overhead of the *signed* letter.

3. Have markers available to add information to the **Algebra Tool Kit**
   chart.

■ Letter #1 from Professor Arbegla

1. Tell students you've received a letter from Professor Arbegla, and read it aloud from the copy you made. As you read, be sure everyone knows the meaning of such words as "malfunctioning." (Have students provide the definitions when possible.)

2. Show the letter transparency on the overhead projector so the class can see it. Focus on the T-table in the letter.

3. Have students discuss with the person beside them what might be going on with the Function Machine. After a few minutes, refocus the class.

4. Pass out the journals and explain that students are to write a draft of a letter to Professor Arbegla in their journals, explaining what they think is going on with the Function Machine. They can use the ideas they discussed in pairs, but each person will write independently.

5. As students are writing, circulate and maintain a quiet environment. Assure students that Professor Arbegla is very interested in their thinking and open to all their ideas.

6. After five–10 minutes, have students close their journals. Ask the class what they think is going on with the machine. Follow their lead in guiding the discourse as they propose answers, and write any solutions on the board.

IN	OUT
25	25
3	3
77	77
48	48
102	102
999	999
5,190	5,190
$\frac{1}{2}$	$\frac{1}{2}$
$\frac{15}{16}$	$\frac{15}{16}$
10.5	10.5

If some students finish before the majority of the class, pose a problem for them to solve.

This sample dialogue can give you an idea of what to expect.

Sample Dialogue:
Why Is the Machine Malfunctioning?

Teacher (TR): Who has an idea about what the machine might be doing?
Tara: It's multiplying by 1.
TR: Tell us more about that idea.
Tara: Well, like you put in the 25 and multiply the 25 by 1 and you get 25.
TR: So, could we write that as a multiplication sentence or equation?
Tara: Yeah, 25 times 1 equals 25. That's how the first number came out of the machine.
TR: Does everyone agree with this idea?

Show of hands.

TR: Do you think it will work for any number that goes in?

All: Yes.

TR and class check the other numbers that went in.

TR: How could we write a number sentence or equation using a letter to represent any number that goes in and gets multiplied by 1?
Donte: Let's say the letter we use is "*n*," then write *n* times 1 equals *n*.
TR: Would you come up and write that for us?

Donte comes up and writes $n \cdot 1 = n$ on the board.

TR: The "*n*" represents any number. It is a *variable*. Will this number sentence be true for ANY number we put in? Talk to your neighbor. See if there are any exceptions.

Provides a moment to discuss.

Maria: What if you put a 0 in?
TR: What do you think about Maria's idea? What do you think will happen if $n = 0$?
Sofia: It'll be 0.
TR: Why?
Sofia: Because 0 times 1 equals 0.
TR: So if we substitute a 0 for *n*, then 0 times 1 equals 0.
TR: Let's take another minute to look at 0. Is there anything special about this number? Talk to your neighbor.

Allows a few moments.

TR: Who'd like to share their thoughts on 0?
Alexa: Well, if you have a number and you add 0 to it, you get the same number.
TR: Alexa, can you come up and give us an example on the board to explain what you mean?

Alexa [as she writes "17 + 0 = 17" on the board]: Well, say you have the number 17 and you add 0. Then you have 17.

TR: Does everyone agree?
TR: Alexa, how could you write that for ANY number?
Alexa: I'm not sure.
TR: Call on someone to help you.

Alexa calls on Angela, who comes to the board.
Angela: Well, let's say we have "*v*" that can be any number. Then we could write $v + 0 = v$.

TR: Would that be true for ANY number? Talk to a neighbor.

Provides a moment.

Ramon: Yes, because you're just adding nothing—and if you don't add anything, then you still have the same number.
TR: I see. You're saying that since the value of 0 equals nothing...the number stays the same. Can anyone think of a number that this wouldn't work for?

No responses.

TR: I think this is something we want to add to our Algebra Tool Kit. Remind me how to write any number plus 0 equals the number.

With class's help, TR records on chart: $v + 0 = v$ for any v.

TR: That was one great idea about 0. Did anyone come up with a different idea?
José: Well, instead of adding, if you times any number by 0, you get 0.
TR: José can you give us an example?
José: Yeah, like 25 times 0 is 0.

TR records on the board: "$25 \cdot 0 = 0$"

TR: So, 25 multiplied by 0 equals 0. Remember, when we "times," we use the word "multiply."
TR: Is this true for any number? For any n, if we multiply it by 0, we always get 0? Any exceptions?

Signs of agreement/no exceptions given.

TR: That's true. Let's add that information to our Algebra Tool Kit. How should we write it?
Cedric: n times 0 equals 0.

TR: Records $n \cdot 0 = 0$ on chart.

TR: Let's go back to where we started when Tara said any number times 1 equals that number. Let's record that on the chart as well.

TR records $n \cdot 1 = n$ on the chart.

7. Ask if there are any other ideas about what the machine might be doing. Continue to have students explain their thinking and record other solutions on the board. Be sure there's agreement and understanding about all the solutions presented.

8. Here's a list of students' explanations of what the function machine was doing to the numbers that went in:

Algebraic Expressions	**Equations**
$n + 0$	$n + 0 = n$
$n - 0$	$n - 0 = n$
$n \cdot 1$	$n \cdot 1 = n$ and $1 \cdot n = n$
$\frac{n}{1}$	$\frac{n}{1} = n$
$n + n - n$	$n + n - n = n$
$n + 999 - 900 - 90 - 9$	$n + 999 - 900 - 90 - 9 = n$
$\frac{(100 \cdot n)}{100}$	$\frac{(100 \cdot n)}{100} = n$
$\frac{[(n - 7 + 7) \cdot 2]}{2}$	$\frac{[(n - 7 + 7) \cdot 2]}{2} = n$

Be aware that the algebraic expressions are not limited! Many creative students have fun making a string of operations that result in the same number exiting the machine.

*By the same token, don't be surprised if your students only come up with one answer to this problem. Over the course of the unit, it's likely they'll discover other solutions in the context of another problem. **Refrain from simply giving them more solutions at this time.***

Dear Mr. Arbegla

I Know what is wrong with the macine. Every time you put a number in the same number comes out. Because the machine is multiplying by 1. Like n x 1 = n its not mest up its just multiplying by 1. From Fabian

Dear Professor Arbegla,
I know what happended to your machine. There is no problem really the machine is just multipling by the number people want to put in for example 100 x 1 = 100. There are other pasibilities to the machine might be adding 0 for example 50 + 0 = 50 One other way would be suptracting zero. There is one last way that I know what might be the answer to what the machine might be doing the one last way is dividing for example 12 ÷ 1 = 12.
from,
Charlie

Dear Professor Arbegla,
Your machine is not realy brocken because our class found out why it is doing it for example: $n = 1 \cdot n$, $n \times 1 = n$, $1 \times n = n$, $n + 1 \cdot n + n$, $n + 8 + 8 - 16 \times 1 = n$, $n \times 1 = n$, $(n \div 1) = n$, $n \div 1 = n$, $n = n$, $(n \times 1) + a - a = n$. I'm havcing fun and I thought math was boring.
Your friend,
Breeanna
Louise

P.S. Your last name is algebra backwords!

Dear Mr. Arbelas,   1·30·01
I know how your Machine works. All that you have to do is times it by one. But if you put in zero it won't come out the same.
— Vanesa

■ To the Tool Kit

1. Check that all the algebraic expressions and/or equations that came out of the discussion were recorded on the **Algebra Tool Kit** chart. Be sure that at least one of the following expressions came out of the discussion:

   $n + 0 = n$

   $n \cdot 1 = n$

2. Reread the expressions and equations with the class. Tell students that **$n + 0 = n$ is a special equation; it algebraically expresses the *identity element for addition*.** Ask if they know what that means. Explain that it's called that because *the number is not changed by adding a zero to it*.

3. Explain that, similarly, **$n \cdot 1 = n$** algebraically expresses the *identity element for multiplication* because the number that's multiplied by one *keeps its value, or "identity," and is not changed* by the operation of multiplication. You may choose to add these expressions and terms to the **Algebra Tool Kit** chart. Have students add any new expressions/equations to the Algebra Tool Kits in their journals.

4. When all ideas have been discussed and recorded, allow time for students to write in their journals a polished letter to Professor Arbegla, explaining what the Function Machine is doing. Tell them to clearly explain one function the machine is performing, giving examples and using algebraic expressions or equations.

5. When students have finished, collect their algebra journals. (You may opt to have students finish the letter at home. In that case, either assign the letter on loose paper and collect the journals now, or have students write in their journals and bring them back for collection at the end of the next session.)

Session 2: Oh, No, Not Again! (Letter #2)

■ What You Need

For the class:

- ❑ the **Fabulous Function Machine** chart from Activity 1
- ❑ the **Algebra Tool Kit** chart from Session 1
- ❑ 1 copy of Professor Arbegla's **Letter #2: Oh, No, Not Again!** (page 39) to be signed
- ❑ 1 overhead transparency of *signed* **Letter #2**
- ❑ an overhead projector

Again, you may choose to create your own letter, posing a problem appropriate for your students to solve independently.

For each student:

- ❑ algebra journal from Session 1
- ❑ 1 *signed* copy of Professor Arbegla's **Letter #2**

■ Getting Ready

1. Copy and trim Professor Arbegla's **Letter #2: Oh, No, Not Again!** (page 39) and have the same colleague sign it.

2. Copy enough *signed* letters for all students.

3. Make an overhead of the *signed* letter.

4. Decide how much time you'll give your students to work on the problem at home, and outline your expectations for this assignment.

 ## ■ Another Letter! Oh My!

1. Tell students you've received another letter from Professor Arbegla, and read it aloud.

2. Show the transparency of the letter, and check that everyone understands what problem the letter poses. Have a student explain it in his own words to the class.

3. Let your students know they're going to take a copy of this letter home, to work on solving the machine's new "malfunction" in their journals. Remind them that the Algebra Tool Kit in their journals can help them. Even if they don't get a solution, you want to hear their ideas and share them with Professor Arbegla. Distribute the journals and a copy of **Letter #2** for students to take home.

4. Outline your expectations for this assignment, and when it's due. Let students know that on the due date, there'll be a class discussion about what the machine is doing.

> There are several solutions to the problem posed in the letter. The most straightforward is $(n \cdot 0) + 7$.

■ Discussing the Letter

1. On the due date of the assignment, have a student reread the problem aloud. Have students share, with a partner or in small groups, their ideas about what's going on with the machine.

2. After a few minutes, ask a student come to the board and create a T-table with the numbers Professor Arbegla put into the machine, as well as the outputs for each number.

3. After the student has returned to his seat, ask another volunteer to share her ideas about what the machine is doing. As she explains, be sure other students are following her thinking. Check to see if others agree, or if they have any questions to ask her.

4. Call on another volunteer to explain his thinking. As necessary, ask questions to help clarify ideas. Refer to the **Algebra Tool Kit** chart if students use the information on it in their explanations.

5. If students are unable to come up with any answers, scaffold the steps to an algebraic expression by asking questions. For example:

- "Look at our **Algebra Tool Kit.** Is there any expression that can help us get started?" [$n + 0 = n$.]

- "What if we let $n = 7$?" Record the equation, $7 + 0 = 7$, and say that the solution to the problem posed in the letter is a two-step function.

- Have students talk to a partner or in small groups after this hint.

- If they're still stymied, provide another guiding question, such as, "What algebraic expression has zero as an answer for any number?" [$n \cdot 0 = 0$.]

IN	OUT
4	7
11	7
5	7
13	7
1	7
0	7
$\frac{1}{2}$	7

Some students have written algebraic expressions for each pair of numbers on the T-table. Though these are usually correct for each pair, they don't work for "any number"(n) that goes into the machine. This is a great time to review that the variable in the algebraic expression must work for all values of "n."

6. When at least one solution has been presented, let students know you'd like them to determine other algebraic expressions that will always result in 7 as further homework. If needed, provide this expression to help guide them: $7 + n - n$.

7. Have students continue working on other solutions as homework or at another designated time. Be sure to allow follow-up time so they can share their solutions.

Students may also come up with these possible solutions:

$7 + (n \cdot 0)$

$\frac{n}{n} + 6$

$n^0 + 6$ (note: $n^0 = 1$)

8. Collect students' journals and have them use loose paper for the follow-up homework, or have them work on solutions in their journals and collect them at the end of the next activity.

Letter #1: Malfunction in the Function Machine

Dear Students,

I know my colleague, your teacher, has shared my Fabulous Function Machine with you. And furthermore, I know you've been able to figure out how it works! That's why I'm writing to you. I really need your help.

It seems someone has been messing around with my machine. It seems to be malfunctioning. Here's the problem: every time I put a number in, the very same number I put in comes out!

For example, here are the last 10 numbers I've put in, and what's come out:

IN	OUT
25	25
3	3
77	77
48	48
102	102
999	999
5,190	5,190
$\frac{1}{2}$	$\frac{1}{2}$
$\frac{15}{16}$	$\frac{15}{16}$
10.5	10.5

No matter what goes in, the same number comes out. At least that's what's happened for the last 453 numbers I've put in!

Your teacher has told me you're very smart. I'd *really* appreciate your assistance with this problem. Please explain to me what MIGHT be going on. I look forward to your reply.

Mathematically yours,

Professor Z. Arbegla

Letter #2: Oh, No, Not Again!

Dear Students,

Thank you very much for your help in solving the first problem with my Fabulous Function Machine. Your teacher shared your ideas and solutions with me. It was very interesting to me to see that there's more than one solution to that problem!

In fact, your ideas helped me with another problem. Over the weekend, I thought my machine was malfunctioning again. This time, whenever I put a number in, a new number came out...but it was the *same number* every time—no matter what went in! Here's an example of what was happening:

IN	OUT
4	7
11	7
5	7
13	7
1	7

Then I decided to try numbers that weren't "counting numbers":

First I tried 0. And guess what? Again, I got 7!

Then I tried a friendly fraction, $\frac{1}{2}$. I still got 7.

No matter what number I put in, the same number—7—came out. Then I called your teacher, and we worked on the problem together and found a rule that worked! We even think it may not be the only one!

Since, according your teacher, you're doing so well with decoding functions and algebraic expressions, I would appreciate your help in finding more possible rules. I look forward to hearing back from you and seeing more solutions to what *seemed* to be a malfunction!

Algebraically yours,

Professor Z. Arbegla

Overview

In this activity, you share your *own* invention—a "Morph Machine" with amazing abilities. When a number enters the "transformer" part of the machine, it's changed, or "morphed," through a series of arithmetic operations. Then the "morphed" number moves from the transformer chamber into the "restorer" chamber of the machine, whereby it converts back to the original number that went in. The "magic," of course, involves algebra. Since the numbers are transformed by a sequence of arithmetic operations, you can use inverse operations to determine numbers the students suggest as input.

In Session 1, students are introduced to the Morph Machine. As they morph numbers, they're using their mental math skills to learn how the machine works. This part of the activity is key to students' success in the next session, when they determine and generalize how the machine operates. As you restore numbers, you create a table to record both the morphed numbers and the original numbers that went in. Using this table, students look for patterns in the morphed output to help them "restore" the original numbers.

In the second session, students are guided to understand more about how the morph machine works. To unravel the numbers that go into the transformer chamber, inverse operations come into play. You guide students in a step-by-step process, first concretely and then algebraically. Students learn how effective algebraic notation can be in expressing mathematical operations, and how simplifying the expression can make "restoring" the numbers a snap.

This activity provides a context for students to hone mental computational skills and to practice writing and simplifying algebraic expressions. They also learn about ***inverse computation operations*** (working backward to solve for an unknown number), which will be crucial in later years as they solve algebraic equations for a variable.

Inverse operations are further described in "Background for the Teacher."

- Inverse operations:

    Addition is the inverse of subtraction

    $$n - 4 = 6$$

    $$n - 4 + 4 = 6 + 4$$

    Subtraction is the inverse of addition

    $$n + 4 = 6$$

    $$n + 4 - 4 = 6 - 4$$

    Multiplication is the inverse of division

    $$\frac{n}{4} = 6$$

    $$\left(\frac{n}{4}\right) \bullet 4 = 6 \bullet 4$$

    Division is the inverse of multiplication

    $$4 \bullet n = 24$$

    $$\frac{(4 \bullet n)}{4} = \frac{24}{4}$$

- Algebra is a tool for writing mathematical statements

- Algebraic notation is a tool for generalizing computation procedures (arithmetic)

Session 1: A New Math Machine

■ What You Need

For the class:
- ❑ 1 sheet of butcher paper approximately 24 in. x 36 in.
- ❑ several wide-tipped colored markers
- ❑ masking tape

For each student:
- ❑ algebra journal from Activity 2

■ Getting Ready

1. On the butcher paper, draw an illustration of the Morph Machine using this graphic as a guide. This is your **Morph Machine** chart for the class.

2. Gather masking tape and the **Morph Machine** chart.

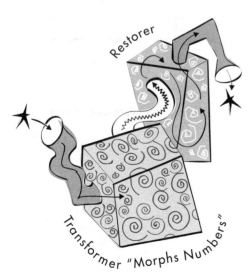

Restorer

Transformer "Morphs Numbers"

■ Introducing the Morph Machine

1. Tell students you were so inspired by Professor Arbegla and her Function Machine that you've created a new math machine of your own—the "Morph Machine." Post your illustration of the machine.

2. Ask what the word "morph" means. After taking several responses, make sure everyone knows that it means to "change" or "transform."

3. Tell students that, as with the Function Machine, numbers get put into your Morph Machine. On entering, they're changed (morphed) in the first chamber, the "transformer," by a series of computations. When the morphed numbers move into the next chamber, the "restorer," your machine lets you determine the original number that went in.

4. Tell students you're going to demonstrate how the machine works. If you collected the journals at the end of the previous activity, redistribute them now. Have each student silently select a number from an age-appropriate domain of numbers and **secretly** write it in their journals.

Third-graders can select whole numbers 1 through 15 or 20, while fifth-graders can typically handle numbers 1 through 50. You know your students' abilities best, so choose what will work for them.

5. Now say you'll put their number into the transformer chamber to be "morphed." Start with the following series of computations. Pause after each step to allow the students to calculate in their heads.

a. Have students **add 1** to their chosen number.   *PAUSE*

b. Have them **double** the result.   *PAUSE*

c. Have them **add 7**.   *PAUSE*

d. Have them **subtract 9**.   *PAUSE*

Though this activity is ideal for strengthening mental math skills, some teachers find that their students need to use paper and pencil to keep track of their calculations.

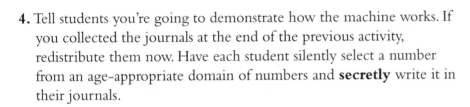

Note: From *your* point of view, the transformation can be represented by the following steps:

Algebraic Expression	Simplest Form
a. $n + 1$	
b. $2(n + 1)$	$= 2n + 2$
c. $2(n + 1) + 7$	$= 2n + 2 + 7 = 2n + 9$

d. $2(n + 1) + 7 - 9$ $= 2n + 9 - 9 = 2n$

To determine the restored (original) number, you need only **divide the morphed number by 2!** If a student gives you the morphed number 22, for example, you'll know the original number was 11.

6. Ask students to begin providing some morphed numbers to restore. As you "restore" their numbers (by dividing by 2), make some movement and sound as if the restorer chamber of the machine is operating.

7. Keep a lively pace. After restoring several numbers, ask if students notice a pattern and can guess their classmates' original numbers for themselves.

Sample Dialogue:
Morphing and Restoring

Teacher (TR): Who has a morphed number?
Sharniece: I got 8.

TR makes a T-table and records 8 on the "Morphed Number" side.

TR: OK, let's put it into the restorer...was your number 4?
Sharniece: Yeah...

TR records 4 on the "Restored Number" side of the T-table.

TR: Who has a different morphed number?
Donte: I got 30.

Morphed Number	Restored Number
8	4

TR records 30.

TR: I think your number was 15.

TR records 15.

Donte: How can you do that?
TR: Let's try another number. Who has one to try?
Julissa: My number is 24.

Morphed Number	Restored Number
8	4
30	15

TR records 24.

TR: Any ideas about what Julissa's number was?
T'era: I think it was 12.

TR: Why do you think that?

T'era: Well, when the first morphed number was 8, it was 4 to start. And that's half of 8. Then when the next number was 30, it was 15. And 15 is half. So, I think 12 because that's half of 24.

TR: Do you all follow T'era's thinking?

Morphed Number	Restored Number
8	4
30	15
24	12

Most agree.
TR records 12.

TR: Does anyone have another idea about what the number that went in could be?

Miki: I'm not sure, but I don't think it will always be half.

TR: Let's try another number and see what happens. Anyone want to give another morphed number?

Raul: 32.

TR records 32.

TR: Talk to a partner about what number you think Raul put into the machine.

Provides time for brief discussion.

TR: Any ideas?

Nico: I think it was a number we couldn't use.

TR: What do you mean?

Nico: I think it was 16 and we were supposed to pick a number 1 through 15.

TR: What do the rest of you think?

Flor: I think 16 too because that's half of 32.

TR: Any other ideas?

In this case, students discovered how to restore the morphed numbers by halving or dividing by two. By looking at the morphed number and the original number, they saw a pattern and a relationship between the two.

■ Morphing New Numbers

1. Tell students you've just reprogrammed the Morph Machine to transform numbers in a new way. Give them the domain of whole numbers to select from, such as a whole number from one through 20 (expressed as $n = \{1–20\}$), and have them silently pick one. Again, they should record the number in their journal and keep it secret!

2. Their numbers are now ready to be morphed. Give students the following sequence of operations, pausing as needed for them to mentally calculate the operations.

    a. **Double the number** (or "multiply by 2").

    b. **Add 2** (to the doubled number).

    c. **Triple the result** (or "multiply by 3").

    d. **Subtract 6.**

Note for your eyes only: These operations can be represented algebraically as follows:

    a. $2n$

    b. $2n + 2$

    c. $3(2n + 2) = 6n + 6$

    d. $6n + 6 - 6 = 6n$

In this case, to determine the students' original numbers, you need to **divide the morphed numbers they give you by 6.** For example, if someone tells you her morphed number is 30, you know the original number was 5.

■ Restoring the Numbers

1. Ask for a morphed number. Make a T-table on the board and record the morphed and restored (original) numbers as in the following example:

Morphed Number	Restored Number
42	7

2. Check with the student that you restored her number correctly.

3. Continue taking morphed numbers from the class and restoring them. Here are a first few numbers given in one fourth-grade class:

Morphed Number	Restored Number
42	7
30	5
36	6
25	—

4. When you've restored and recorded about eight of the students' numbers, ask how they think the machine is operating to restore their numbers. Ask such questions as "What do you think is happening in the restorer chamber?" "Can you figure out what the restorer did to your morphed number?"

5. Provide time for students to share their ideas with a neighbor or in small groups. Ask them to record in their journals what the restorer is doing, but don't debrief their ideas in class. (You can assess their understanding when you collect their journals.)

■ More Morphing

1. Time permitting, select a new series of operations for your students to use to morph a new number.

2. Give them a range of numbers to choose from and call on a student for a morphed number. Restore it and record the numbers as before.

3. Collect the journals.

Session 2: What's the Magic in the Morph?

■ What You Need

For the class:
❏ the **Morph Machine** chart from Session 1
❏ the **Algebra Tool Kit** chart from Activity 2
❏ several wide-tipped colored markers

For each student:
❏ algebra journal from Session 1

*Sometimes a student will make a calculation error. In one instance, a student offered the number 25. Since 25 isn't evenly divisible by 6, the teacher knew an error had occurred. She reminded students that the machine could only process whole numbers, and pointed out that there must have been a malfunction somewhere, because the number could not be restored. Should this happen in your class, give the student a chance to transform the number again, but **don't dwell on the error or try to fix it in that moment,** so as not to disrupt the overall flow of the activity.*

■ A New Way to Record

1. Open this session by asking students what they've learned so far about your Morph Machine. Listen to their ideas and build on them as appropriate as you introduce them to a new recording system and inverse operations. Pass out the journals.

2. Tell students that the Morph Machine is ready to transform numbers again. Write the following on the board. Leave a large space between the "*n*" and the "Morphed *n*" lines, as follows:

$n =$ _____

Morphed $n =$ _____

3. Tell students that *n* represents a number from one through 20; $n = \{1–20\}$. Have them silently select an *n* to go into the transformer chamber and secretly record it in their journals.

4. Give them the following calculations to apply to their numbers:

a. **Add 1 to your number.**

b. **Triple the result** (or "multiply by 3").

c. **Add 4.**

d. **Subtract 7.**

5. Ask a student to give you her morphed number. Record her number (say, 15) on the board as the "Morphed *n*" number. Put a question mark in the space for *n*, to reinforce that it's the original number you want to determine (restore).

$n =$?

Morphed $n =$ 15

6. Have students try to restore the original number as you "process" the morphed number in the restorer chamber. Record the restored number (5, in our example) in place of the question mark.

$n =$ 5

Morphed $n =$ 15

7. Ask if anyone else thought the original number was 5. Have students talk with a partner or in small groups about how the restorer may be decoding the number.

8. Ask for their ideas. After listening to their thinking, ask them to recall the operations they performed on their original number.

9. Record the computation steps on the board as students describe them. Guide students to either work backwards from the last step, "Subtract 7," or start at the first computation, "Add 1," so the steps are sequential.

$n =$	5
Add 1	+ 1
Multiply by 3	x 3
Add 4	+ 4
Subtract 7	− 7
Morphed $n =$	15

> *This is an example of using "x" rather than "•" to indicate multiplication; it clearly states the operation being used for calculation purposes.*

10. After all steps are recorded, have students go through each step to see that if $n = 5$, the Morphed $n = 15$.

■ A Closer Look at the Restorer Chamber

1. Erase the 15 and the 5 but leave the computation steps on the board. Ask for a different morphed number (say, 21) and record it as the "Morphed n" as follows:

$n =$	
Add 1	+ 1
Multiply by 3	x 3
Add 4	+ 4
Subtract 7	− 7
Morphed $n =$	21

2. Challenge students to determine the original number. Circulate as they work.

3. Focus the class and ask someone to try to identify the original number. Ask how she arrived at that number, and have her use the board to make the presentation as clear as possible. Have other students assist as necessary and ask questions to guide them.

4. If your students are struggling with a way to explain, guide them through the process. Start by focusing on the morphed number (21). Ask what 21 was the step before. [28.] How did they figure out that number?

5. Show the step on the board:

$n =$	_____	
Add 1	+ 1	
Multiply by 3	x 3	
Add 4	+ 4	
Subtract 7	− 7	*Before 7 was subtracted to get 21, the number was 28. To determine that, you have to **add** 7 to 21.*
Morphed $n =$	21	

6. Continue to work backwards, explaining each step. Encourage students to assist and explain as well.

$n =$	_____	
Add 1	+ 1	*Subtract 1 from 8. You get 7— the original number that went into the transformer!*
Multiply by 3	x 3	*Divide 24 by 3. [8.]*
Add 4	+ 4	*Subtract 4 from 28. [24.]*
Subtract 7	− 7	*Before 7 was subtracted to get 21, the number was 28. To determine that, you have to **add** 7 to 21.*
Morphed $n =$	21	

Work backwards to restore

7. Let students know that they're using an important strategy in problem solving—**going backwards.** They're also using opposite, or *inverse,* operations. Tell them the "magic" in the morph machine is ALGEBRA—and they're using it!

8. Say the morphed number was 21. Go back to the computational operation before that—"subtract 7." That tells you that some number (*n*) minus 7 equals 21.

9. Write that statement in algebraic form: $n - 7 = 21$. To solve for *n*, you need to use the *inverse operation of subtraction* and **add** 7 to both sides of the equation.

$$n - 7 = 21$$
$$n - 7 + 7 = 21 + 7$$

> • If you perform an operation to one side of the equation, the same operation must be performed to the other side, to maintain the equality.
>
> • Addition is the inverse of subtraction.

$$n + 0 = 28$$
$$n = 28$$

10. Depending on your students' abilities, you may want them to write the algebraic statements for each of the prior computational steps.

11. Ask for another student's morphed number (we've used 33). Have students use their journals to determine the volunteer's original number using inverse operations.

12. After ample time, have a student come to the board to explain to the class how to restore the number.

$n =$	?	
Add 1	+ 1	$12 - 1 = 11$
Multiply by 3	x 3	$36 \div 3 = 12$
Add 4	+ 4	$40 - 4 = 36$
Subtract 7	− 7	$33 + 7 = 40$
Morphed $n =$	33	

Work backwards

13. For older or more experienced students, you can show the algebraic way to write each arithmetic step, as well as how to simplify the algebraic expression:

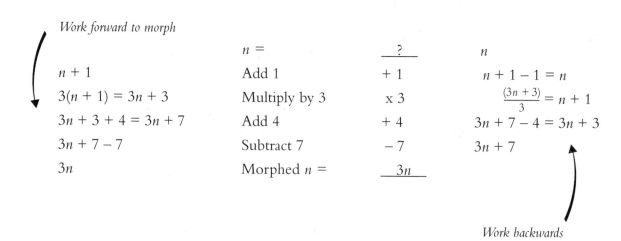

Work forward to morph

	$n =$	?	n
$n + 1$	Add 1	$+ 1$	$n + 1 - 1 = n$
$3(n + 1) = 3n + 3$	Multiply by 3	x 3	$\frac{(3n + 3)}{3} = n + 1$
$3n + 3 + 4 = 3n + 7$	Add 4	$+ 4$	$3n + 7 - 4 = 3n + 3$
$3n + 7 - 7$	Subtract 7	$- 7$	$3n + 7$
$3n$	Morphed $n =$	$3n$	

Work backwards

Therefore, to determine any n, divide the morphed number by 3. The other steps just make it *seem* complicated to restore numbers!

■ More Morphing and Restoring

We recommend that when you do this for the first time, you use **addition** and **multiplication** as the computational operations. Subtraction can take students into the realm of negative numbers, and division can create fractions that become challenging.

1. "Reprogram" the Morph Machine to do another series of operations. Give students the computational steps to morph their numbers. When a student shares his morphed number with the class, have students restore (try to determine) the original number. Use algebraic expressions according to your students' skills and abilities.

2. Ask students if they want to add anything from this lesson to the **Algebra Tool Kit.** Whatever you record on the class chart, be sure that they add it to their journal Tool Kits too.

3. For homework, have students create a process to morph and restore numbers. (Allow time in the next session for students to present their machines in class.) This provides computational practice, reinforces inverse operations, and uses algebraic thinking.

4. Collect students' journals and have them use loose paper for the follow-up homework, or have them work on solutions in their journals and collect them at the end of the next activity.

■ Going Further

To provide a challenge, have students create their own morph machines and ask their classmates to determine the operations that morph the numbers that go in.

■ Special Going Further

1. More Morphs

On pages 55–56, we've included two more morph processes to challenge your students. These are the kind of morphs that really seem magical, as the original number appears within the morphed number. Try them out first to be sure your students have the skills to do them. After students have done both successfully, encourage them to design similar morphs.

2. The Art of Morphing Shapes

Not only can numbers be morphed, but shapes, too, can be changed in dramatic ways. The Dutch artist M.C. Escher was a master at morphing (changing) geometric shapes. In many cases, the morphed shape would "tessellate" (form into a mosaic) in a plane. Imagine an altered square made into a tile piece. Using many of these very precise tiles, the area of a floor or counter could be covered. The tiles would fit together smoothly, with no gaps, and overlays wouldn't be needed to cover the area.

Here's an example of how to alter a square.

Start by distributing squares made of card stock. (Pre-shade them, if you wish; see #1.) Give students the following instructions:

1. Shade the front of the square to distinguish it from the back.

2. Cut a piece (angular or curvy) from **one side** of the square. (Caution students not to cut too close to the opposite edge.)

For more information on tessellations, consult pages 146–147 and 149 of the GEMS Build It! Festival *teacher's guide*. This "Going Further" activity on morphing a square is from that guide.

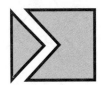

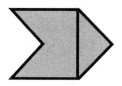

3. Slide the cut-out piece to the opposite side and tape it on, shaded side facing up.

4. Repeat the procedure, cutting a shape out of one of the two remaining **unaltered** sides. (Take care not to cut into the piece you attached!) This shape should be different from the first one you cut out.

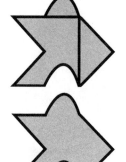

5. Slide the new cut-out to the opposite side and tape it down—again, shaded side up.

6. Look at your new overall shape and see what it looks like! Rotate the whole thing to inspire different possibilities. (For instance, maybe it looks like a fish in one direction, a frog in another.)

7. Allow a few minutes for students to experiment. Depending on your students' abilities, you can assess their prior knowledge of two related functions they'll be introduced to later in the unit: **_area_** and **_perimeter._** Ask your class this question:

- Since you used the entire square (and only the square) to create your shape, the area of the new shape remains the same as the original square. Does the perimeter of the new shape also remain the same? (Hint: A string may help you check the perimeter of the new shape.)

8. For an optional and fun finale to this activity, students can repeatedly trace their shapes onto paper until their tessellations take on a whole new life!

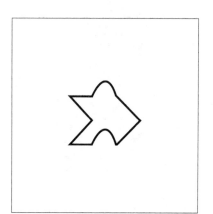

More Morphs

Favorite Number Morph

The following series of calculations allows your students to morph a favorite number between 1 and 100. The morphed number will be the original number multiplied by 10—so it's easy to "guess" their favorite numbers by simply dividing by 10. Here's the scenario you can use with your class:

> Start with your favorite number
>
> Add 5 to the number
>
> Multiply the sum by 2
>
> Add 3 to the product
>
> Multiply the sum by 5
>
> Subtract 65 from the product

For example, if the favorite number was 23, here are the calculation steps to morph that number:

		Algebraic Expression
Start with your favorite number:	23	n
Add 5 to the number:	28	$n + 5$
Multiply the sum by 2:	56	$2(n + 5) = 2n + 10$
Add 3 to the product:	59	$2n + 10 + 3 = 2n + 13$
Multiply the sum by 5:	295	$5(2n + 13) = 10n + 65$
Subtract 65 from the product:	230	$10n + 65 - 65 = 10n$

It is likely to amaze your students that their favorite numbers appear in the morphed number! You'll be able to guess their favorite numbers with ease—as long as the calculations are correct! Simply divide by 10.

After you've guessed several numbers, ask students how their morphed number is related to their favorite number. [It's 10 times larger.]

Have students try this Favorite Number Morph process on at least two other people. Challenge them to determine how the morphed number grew to a magnitude 10 times its original size. Algebraic expressions will be helpful!

Magical Age Morph

The following series of calculations allows your students to morph their ages. The morphed number will be their age multiplied by 100, so it's easy to guess their age by simply dividing by 100.

> Start with your age
>
> Multiply your age by 5
>
> Add 3 to the product
>
> Multiply the sum by 4
>
> Add 13 to the number
>
> Multiply the sum by 5
>
> Subtract 125 from the product

For example, if the student's age is 11, the calculations would be as follows:

		Algebraic Expression
Start with your age:	11	n
Multiply your age by 5:	55	$5n$
Add 3 to the product:	58	$5n + 3$
Multiply the sum by 4:	232	$4(5n + 3) = 20n + 12$
Add 13 to the number:	245	$20n + 12 + 13 = 20n + 25$
Multiply the sum by 5:	1225	$5(20n + 25) = 100n + 125$
Subtract 125 from the product:	1100	$100n + 125 - 125 = 100n$

Again, it'll seem quite amazing for students to have their age appear as part of the morphed number! Reminder: to guess the age, divide by 100.

After you've guessed several numbers, ask students how their morphed number is related to their age. [It's 100 times larger.]

Have students try this Magical Age Morph process on at least two other people. Challenge them to determine how the morphed number grew to a magnitude 100 times its original size. Algebraic expressions will be helpful!

Overview

In this activity, students are introduced to one of Professor Arbegla's colleagues, Professor Lina LaBarge. Professor LaBarge works in the physics department and has designed a scale. She experiments mixing and matching different weights on her scale until it's perfectly balanced.

After students learn how Professor LaBarge's scale works, they're given a problem from her to solve. They're asked to determine the weights of two different objects—triangular and square weights. This problem introduces students to the scale as a metaphor for equations. Students determine the weights to balance the scale and record equations using variables. As they solve additional problems, students encounter cases in which there's only one solution, and problems that can be generalized and expressed to obtain an infinite number of solutions.

As they balance the scales, students use T-tables and look for patterns. They also write equations for the weights using numbers and variables. In addition, these balance problems provide a context for discussing the **commutative property of addition,** which is added to the **Algebra Tool Kit** chart and the students' journals.

Equations and the commutative property of addition are further described in "Background for the Teacher."

Algebra Tools and Key Concepts— Activity 4

- A "property" is a computational rule for numbers

- Commutative property of addition: $a + b = b + a$

- Writing algebraic expressions

- Solving algebraic equations for a variable

■ What You Need

For the class:
- ❑ the **Algebra Tool Kit** chart from Activity 3
- ❑ several wide-tipped colored markers

For each student:
- ❑ algebra journal from Activity 3
- ❑ *(optional)* $\frac{1}{4}$ cup dry beans (black, pinto, or any small bean)
- ❑ *(optional)* 1 container for beans (8- or 16-oz. yogurt, sour cream, and/or cottage cheese containers work well)

Using beans to represent numbers (pounds) may seem very simple for your students, but it lays the foundation for upcoming problems. It also ensures success for all.

■ Getting Ready

Decide if you'll use beans for this activity. If you do, fill each plastic container with approximately $\frac{1}{4}$ cup of beans. Set aside two containers per pair of students.

GO Session 1: Professor LaBarge's Scale

1. As you're passing out the journals, introduce a colleague of Professor Arbegla's: Professor Lina LaBarge, who works in the physics department. Tell the class Professor LaBarge has designed a scale.

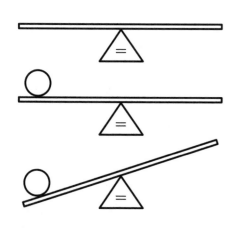

2. Draw a picture of the scale on the board. Call students' attention to the "equal sign." Ask if anyone has ever used a scale, and how it works.

3. Tell students that Professor LaBarge has a variety of weights for her scale. Draw a circle on one side of the scale to represent one weight.

4. Ask what would happen if this was a real weight? Agree that the scale would tip to that side.

5. Ask how the scale could be balanced. Be sure students know that to balance, weights on each side of the scale need to be equal.

6. If you're using beans with this activity, hand each pair two containers.

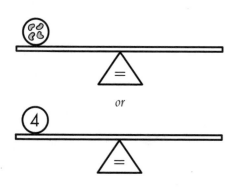

or

7. Have students draw a level scale in their journals (like your model on the board) and add a circle to one side. Tell them the circular weight weighs four pounds. If you're using beans, have students place four beans in the circle to represent the four pounds. If not, have them write "4" in the circle.

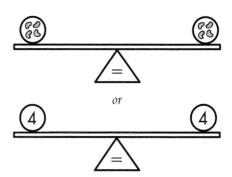

8. Ask the students to balance their scales. Have one student share his solution. Be sure everyone has drawn a second circle of the same size on the other side of the scale. Check that each student has four beans or a "4" in the second circle.

■ Solving Professor LaBarge's Problem

1. Tell students that Professor LaBarge was using her scale, yesterday. She put a square weight on one side of the scale and had to use three triangular weights to balance it. Unfortunately, she forgot to label the weights!

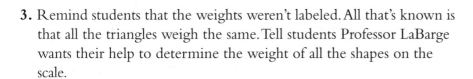

2. On the board, draw a level scale with a square on one side and three triangles on the other. Have student draw this in their journals.

3. Remind students that the weights weren't labeled. All that's known is that all the triangles weigh the same. Tell students Professor LaBarge wants their help to determine the weight of all the shapes on the scale.

4. Tell students they're going to figure out the possible weights of the triangle and the square. Draw a triangle and a square on the board as shown, and have students begin working on the problem.

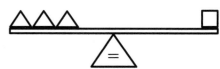

5. Circulate as they work. As students get one solution, encourage them to look for others. When most students have at least one solution, focus the class to discuss their results.

6. On the board, make a T-table to record the weight of the triangles and square as follows:

Recording the weights in this way will help students see a relationship, or pattern, between the triangular and square weights.

7. It's likely a student will say that the triangles weigh one pound each and the square weighs three pounds. Ask if other students agree. How can they prove that the scale will balance if this is true?

△ | □
1 | 3   1 + 1 + 1 = 3

8. Write the equation that expresses the equality, or balance, of the scale: 1 + 1 + 1 = 3. Record that result on your T-table.

9. Continue asking for other possible relationships between the square and triangular weights. Check that students agree with each solution, record the equations that balance the scale, and add to the T-table. For example:

△ | □
1 | 3   1 + 1 + 1 = 3
5 | 15   5 + 5 + 5 = 15
2 | 6   2 + 2 + 2 = 6

10. After enough numbers have been listed for a pattern to emerge, ask students if they could determine the weight of the square, given the weight of only one triangle. For example, if the triangular weight was $\frac{1}{2}$-ounce, how much would the square weigh? [$1\frac{1}{2}$ ounces.]

11. Ask a volunteer to explain how she determined the weight of the square. Be sure all students understand that the weight of the square is three times the weight of one triangle.

3 △ = 1 □

△ | □
1 | 3
5 | 15
2 | 6
$\frac{1}{2}$ | $1\frac{1}{2}$
⋮ | ⋮
t | 3t

12. Let the weight of the triangle be represented by the variable *t*. Ask students how to represent the weight of the square [$3t$.], and record it on the T-table. Below the T-table add the algebraic sentence for this: $3t$ = the weight of the square.

13. Tell students that algebra allows us to ***generalize*** the relationship, or pattern, between the triangular weight and the square weight.

14. *For fourth- and fifth-grade students,* ask how, if they were given the weight of the *square,* they could determine the weight of one triangle. [Divide the weight of the square by three.]

■ What Weight Will Balance?

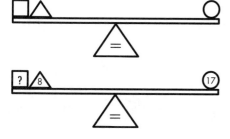

1. Draw another scale, this time with a triangle and a square on one side and a circle on the other.

2. Tell students that you know the circle weighs 17 ounces and the triangle weighs eight ounces. Record the numbers inside the shapes on the scale, and put a question mark inside the box.

3. Draw each shape again and give its value, as shown.

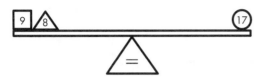

4. Ask how much the square needs to weigh to balance the scale. It's likely your students will rapidly tell you the square weighs nine ounces. Record "9" in the square.

5. Record the equation that the scale originally represented, $n + 8 = 17$. Ask a volunteer to explain how he determined the weight. Many students are likely to say that they know $9 + 8 = 17$.

6. To reinforce your students' understanding of inverse operations to solve for a variable in an equation, go through the following steps to show how to solve this problem algebraically.

$n + 8 = 17$
$n + 8 - 8 = 17 - 8$

> • If you perform an operation to one side of the equation, the same operation must be performed to the other side, to maintain the equality.
>
> • Subtraction is the inverse of addition.

$n + 0 = 9$
$n = 9$

■ Commutative Property of Addition

1. Tell students that Professor LaBarge also wants to find other combinations of **two different weights that could balance her 17-ounce circular weight.**

2. Let students know that Weight #1 will be labeled with the variable n, and Weight #2 will be labeled with the variable w. Record those letters in a new T-table:

n = Weight #1 | w = Weight #2

3. Write the algebraic equation for the sum of the weights on the board:

$$17 = n + w \text{ and } n + w = 17$$

Ask volunteers to "read" the equations and explain in words what the equations are stating.

4. Ask students to determine some possible weights to balance the circular weight. Have them find as many solutions as they can and record them in their journals.

Depending on your students' number sense, they may present solutions using fractions or decimals (zero is not possible in this case).

5. Circulate as they work. When most students have several answers, provide time for them to share their solutions with each other. Ask for and record their solutions in the T-table on the board. Here's a list of solutions your students should be able to generate:

n = Weight #1	w = Weight #2
1	16
2	15
3	14
4	13
5	12
6	11
7	10
10	7
11	6
12	5
13	4
14	3
15	2
16	1

Addends are described in "Background for the Teacher."

6. As students give answers, they're likely to notice that there are two answers using the same addends for 17, to balance the circular weight. For example, Weight #1 could weigh 10 ounces and Weight #2 could weigh seven ounces, or Weight #1 could weigh seven ounces and Weight #2 could weigh 10 ounces. If students haven't noticed, point it out and record those two equations:

$$10 + 7 = 17$$
$$7 + 10 = 17$$

7. Point out that regardless of the order in which the 7 and the 10 are added, the answer is 17. Look at the other similar pairs of numbers in the table, such as 12 and 5 or 1 and 16. Observe that their sums also equal 17, regardless of the order in which they're added.

8. Tell students this illustrates another important property of our number system. Refer back to the two equations you recorded:

$$10 + 7 = 17$$
$$7 + 10 = 17$$

Since they both equal 17, they can also be written:

$$10 + 7 = 7 + 10$$

Although you won't bring this up with your students, this is the Transitive Property, or relation; if a = b and b = c, then a = c.

9. Say that this property has a special name: the ***commutative property of addition.*** It applies for *any* two numbers, and can be written algebraically as:

$$a + b = b + a$$

10. Have students explain the property in their own words to a neighbor, to reinforce the meaning of commutativity. Add the commutative property for addition to the **Algebra Tool Kit,** and have students enter it into their journals. Then collect the journals.

Session 2: More Balancing Acts

In this session, you'll pose additional problems to help students understand that **equations—like the scale—must be equal on each side of the equals sign.** Pass out the journals before you pose the problems to your class, and collect them after students have finished with their final problem.

> ### Alge-Branch Point for Teachers
>
> Select problems below (or create your own) that you feel are appropriate for your class to solve. Have students present the solutions and encourage the use of algebraic notation.

■ Problem #1

1. On the board, draw a scale with three weights: on one side draw a square that weighs more than eight pounds and a triangle that does not weigh eight pounds; on the other side draw a circle that weighs 17 pounds. List their values beside the scale, using both shape symbols and variables.

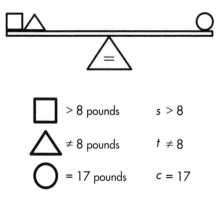

2. Be sure the students are clear on the restrictions for each weight. Have student pairs explain to each other what they know about the weights and what they need to determine to balance the scale. Have them find the possible values for the square and the triangle, recording their work in their journals.

3. Provide beans as needed and circulate as students work.

4. After several pairs seem to have determined several solutions, refocus the class. Ask for values (for the square and triangle) that solve the equation. Check that the rest of the class agrees. Record the solutions in a T-table on the board.

5. Ask for more possible solutions, checking that other students agree with the answers. This is likely to spark discussion, especially if some volunteered solutions are inaccurate. Pose questions to clarify student understanding. Add any additional answers to the T-table.

Here's a list of possible solutions using whole numbers:

□ = s	△ = t	◯ = s + t
10	7	17
11	6	17
12	5	17
13	4	17
14	3	17
15	2	17
16	1	17

Students are likely to observe that the more restrictions on the square and triangle weights, the fewer solutions there are.

■ Problem #2

1. On the board, draw a scale with four triangles on one side and a hexagon on the other.

2. Have students or student pairs determine possible weights for the triangles and the hexagon. Ask them to come up with at least five possible solutions, and have them record their work in their journals.

3. After students have had time to work, ask for and record their solutions on a new T-table on the board.

4. Ask students what the hexagon would weigh if each triangle weighed 71 grams. Ideally, they'll be able to calculate 284 grams. Ask a volunteer how she determined the answer.

△ = t	◯ = h
1	4
2	8
9	36
5	20
1,000	4,000

5. Ask how to generalize the relationship of the weight of the hexagon to the weight of the triangles. Guide them to understand that since there are four triangles, the weight of the hexagon will be four times that of a single triangle.

6. Using t to represent the weight of one triangle, ask a student how to represent the weight of the hexagon in relation to the weight of the triangle. Record on the T-table as follows:

$\triangle = t$	$\hexagon = h$
1	4
2	8
9	36
5	20
1,000	4,000
⋮	⋮
t	$4t$

7. Record this relationship in equation form: $4t = h$. Have students explain in words what this equation states symbolically.

■ Problem #3

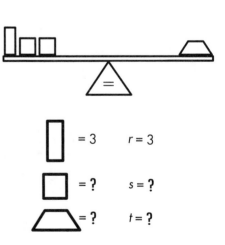

= 3 $r = 3$

= ? $s = ?$

= ? $t = ?$

1. On the board, draw a scale with one rectangle and two squares on one side and a trapezoid on the other. The only known value is the rectangle's weight: three pounds. List their values beside the scale, using both shape symbols and variables.

2. Have a volunteer explain what weight is known and what weights need to be determined to balance the scale.

3. Ask how to write an equation using variables and record examples on the board. Here are some responses you may hear:

$r + s + s = t$
$s + s + r = t$
$r + 2s = t$
$2s + r = t$
$t = s + r + s$

Have volunteers "read" the equations and explain in words what the equations are stating.

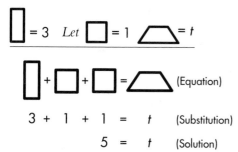

$\rlap{\rule{0.6em}{1.4em}} = 3$ *Let* $\square = 1$ $\diagup\!\!\overline{}\!\!\diagdown = t$

$\rule{0.6em}{1.4em} + \square + \square = \diagup\!\!\overline{}\!\!\diagdown$ (Equation)

$3 + 1 + 1 = t$ (Substitution)

$5 = t$ (Solution)

4. Guide the class through one solution. The given information is the weight of the rectangle—three pounds. Assign the squares a weight of one pound each. Substitute the values for the rectangle and squares into the equation. Finally, solve for the value of the trapezoid's weight. [5 pounds.]

5. Ask if this is the only solution. Why or why not? Once it's agreed that more solutions can be found, remind them that *only the rectangle* has a specific weight of three pounds, and it cannot change.

6. Have students work in pairs to find values for the squares and the trapezoid. Remind them that the squares are the same weight.

7. Provide time for students to find the values. Circulate as they work and encourage them to find more than one solution.

8. Have students present their solutions to the class and ask the other students to check the accuracy of their ideas. (Be sure both squares have been assigned the same weight in any given solution!)

9. Ask students if they can predict whether the weight of the trapezoid will be an even or odd number. Have them explain their reasoning.

Note for teacher: If your students are having difficulty explaining whether the sum will be even or odd, this is one way to explain the outcome to them. Depending upon their skills and abilities, you can decide whether or not to share this information. The sum of the two squares (*s*) will always be an **even** number, because after you double the weight of one square, the result can be divided evenly. Since the rectangle's weight is constant at 3, an **odd** number, and the sum of the two squares is an **even** number, the weight of the trapezoid will be odd. This can be represented this way:

Doubling a number results in an EVEN number:
2 • *even* = even + even = even
2 • *odd* = odd + odd = even

An even number plus an odd number results in an ODD number:
even + odd = odd
odd + even = odd

In the case of this equation, "rectangle + two squares = trapezoid," the result is:

$3 + (s + s) = t$

odd + even = t

odd + even = odd

Therefore the trapezoid's weight is odd.

■ Going Further

1. Yet More Balancing Acts

On a regular basis throughout the unit, pose other balance problems for students to solve, to provide additional practice in writing and solving equations. Be sure to include some problems with additional challenges, such as using fractions and decimals and putting restrictions on the variables (weights).

2. Number Sense Riddle

Read the book *One Riddle, One Answer* to your class (see "Resources," page 130). In this brief story, a Sultan's daughter—who loves numbers and riddles—creates a riddle using facts about numbers and operations. Any young man who hopes to be her husband must first solve the riddle! See if your class can use their number sense to come up with the solution before it's revealed in the story.

3. Creating Equations with Mobiles

The assessment activity on page 126 also serves as an excellent "Going Further" for Activity 4, as students use mobiles to create equations with variables.

Overview

Through another excited letter from Professor Arbegla, your students are introduced to the distributive property as the professor's "great multiplication discovery."

After the letter is read in Session 1, students use concrete materials to gain an understanding of the ***distributive property over addition***, as you provide opportunities to apply this property to problems appropriate to their skills and abilities. As part of the class discussion, you'll record equations in a step-by-step sequence to connect the concrete representations to the numerical representations. This prepares students for using algebra—an abstract representation—to generalize the distributive property.

In Session 2, students discover that there's also a ***distributive property over subtraction.*** They apply this law to solve more multiplication problems, and write the algebraic expression to generalize the distributive property. Finally, they apply what they've learned about the distributive property as they decide how to solve other multiplication problems.

Students continue to have opportunities to practice the distributive properties through "warm-up" and homework problems, and as a "sponge" activity.

The purpose of this activity is to familiarize students with one of the important laws of operations. Though they'll see the distributive property expressed algebraically, it's most important that they have opportunities to **apply it with numbers.** This will lay the foundation for deeper exploration in later mathematics courses.

Sponge Activities

Often, during the academic day, you're presented with a few extra moments you can take advantage of—such as before recess or lunch. These times are perfect for posing problems, including these multiplication problems.

■ What You Need

For the class:
- ❑ the **Algebra Tool Kit** chart from Activity 4
- ❑ 1 copy of Professor Arbegla's **Letter #3: My Multiplication Discovery** (page 84) to be signed
- ❑ several wide-tipped colored markers
- ❑ *(optional)* 1 overhead transparency of *signed* **Letter #3**
- ❑ *(optional)* an overhead projector

For each pair of students:

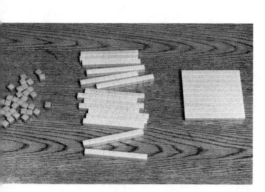

- ❑ 1 set of base ten blocks★ (wooden, plastic, or paper) in the following approximate denominations (the quantity will vary depending on the magnitude of the numbers used in the activity):
    - ___ 35 ones (units)
    - ___ 15 tens (longs)
    - ___ 1 hundred (flat)
- ❑ 1 bag or container to hold the block set

For each student:
- ❑ algebra journal from Activity 4

★You can order sets of base ten blocks (see "Resources" on page 130) or make paper models from the master on page 85.

■ Getting Ready

1. Familiarize yourself with Professor Arbegla's **Letter #3: My Multiplication Discovery** (page 84) and make any changes or additions you think best for your class. Copy and trim it, and have it signed with the same signature as on previous letters.

2. If you wish, make an overhead transparency of the *signed* letter so your students can read along with you.

3. For each student pair, gather base ten blocks or make paper models (page 85; by cutting down the center of the heavy lines, the blocks will retain their borders). **Students will use the blocks twice:** once for the explanation of the "trick," and once to calculate "4 • 38."

4. Read through "Background for the Teacher" (page 107) for additional information on the distributive property.

5. Gather the letter and markers.

Session 1:
The Professor's Multiplication Discovery

1. As you distribute the journals, tell the class you recently met with Professor Arbegla, and you've never seen her so excited! You thought perhaps she'd invented another machine, but that wasn't it.

2. Explain that she loves math and knows lots of interesting things about it, but she'd always had trouble with multiplication of numbers that weren't part of the "times table." She found herself taking a very long time and sometimes losing track of numbers, especially when she needed to do multiplication quickly or in her head.

3. Emphasize that, of course, the professor does know her "times table" and can do multiplication. But somehow, deep down, she felt there had to be an easier way for certain kinds of problems. Recently, she discovered an incredible trick—and when she explained it to you, you asked her to write a note about it to your class.

4. Read **Letter #3: My Multiplication Discovery** aloud to the class. (If you like, show the overhead of the letter as you read.)

■ Arbegla's Multiplication "Trick"

1. Tell students you think Professor Arbegla's math trick is awesome, and—if it turns out that it really works all the time—you wish you'd known about it when you were in school!

2. Explain that the best way to describe the trick is to use an example. Ask students to *mentally* calculate the product of two numbers—a single digit number and a two-digit number. Depending on their skills, choose numbers that will be challenging but not discouraging. Here are suggestions by grade level:

Grade 3:	4 • 26
Grade 4:	6 • 59
Grade 5:	9 • 87

3. After a short time, ask for their answers. **Don't be surprised if this is challenging and some students can't come up with an answer!**

4. Tell them you're very interested in *how* they got their answers. As you listen to their reasoning, build on any ideas that are helpful in explaining the next steps.

■ Using Base Ten Blocks to Understand the "Trick"

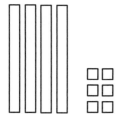

1. Tell students that to help them understand how Professor Arbegla's trick works, they'll use the math tool called "base ten blocks." Distribute the sets of blocks to student pairs.

2. Review how the blocks work. As a warm-up, ask them to build a few numbers. (Start with 35, for example, and then change the number to 53.)

3. Ask student pairs to build the number 46. [Four tens and six ones.] Write 3 • 46 on the board and say you want to know what it equals.

4. Ask what three times 46 means. Be sure students know it's the same as 46 + 46 + 46. To determine how much it equals, they'll need to build the number 46 two more times.

5. Have pairs build 46 twice more. They'll have the following:

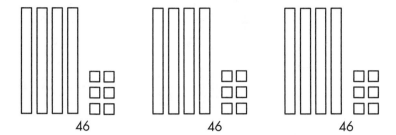

46   46   46

6. Next, ask students to determine what 3 • 46 equals, using the blocks. They should be able to explain how they got their answers, which should reflect the following reasoning (ask guiding questions as needed):

a. Group the tens from each 46 built; there'll be 12 tens.

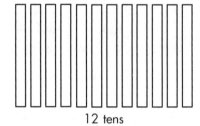

12 tens

b. Exchange 10 tens for a hundred; there'll be two tens left.

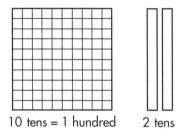

10 tens = 1 hundred   2 tens

c. Group the ones (18 total) and exchange 10 ones for a ten; there'll be eight ones left.

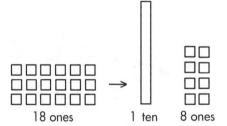

18 ones   1 ten   8 ones

d. Finally, group all the tens (there will be three) and read the number—138.

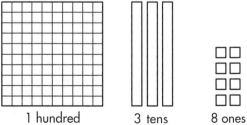

1 hundred   3 tens   8 ones

You can start by grouping the ones first. Then there'll be 13 tens to regroup. After you exchange 10 tens for a hundred, you'll have three tens left.

7. Next, ask how to write the number 46 using tens and ones. Record 40 + 6 on the board. In the problem 3 • 46, there are three 46s. That means there'll be three 40s and three 6s. Ask students to build three 40s and three 6s.

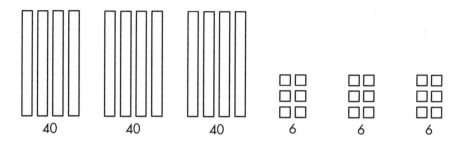

40   40   40   6   6   6

8. Ask students to find the sum of the all the blocks—the answer, of course, is again 138. Some students may think you're a bit zany to ask this question, but others may still have to calculate the answer!

9. Tell students you had them do this to demonstrate Professor Arbegla's trick—which is called the **distributive property over addition.** Because of this property, the answer is the same regardless of how they calculated it.

10. Record what they did with the blocks on the board as follows:

Build 46 three times   =   Build 40 and 6 three times each

‖‖∷ ‖‖∷ ‖‖∷ = ‖‖ ‖‖ ‖‖ ∷ ∷ ∷

46 + 46 + 46   =   40 + 40 + 40 + 6 + 6 + 6

11. As you write the sequence of equations on the board, have students regroup their blocks. This connects the concrete representation to the numeric representation.

$$
\begin{aligned}
3 \bullet 46 &= 3 \bullet (40 + 6) \\
46 + 46 + 46 &= (3 \bullet 40) + (3 \bullet 6) \\
138 &= 120 + 18 \\
138 &= 100 + 20 + 10 + 8 \quad \text{(hundreds + tens + ones)} \\
138 &= 138
\end{aligned}
$$

12. Give students another problem, such as 4 • 38, to solve using the blocks. As students work, circulate and observe how they're approaching the problem.

13. When the students are done, have volunteers explain how they solved the problem. Be sure both methods of solving are shown: 38 + 38 +

38 + 38 **and** 4 • (30 + 8). Review the steps by recording the sequence of equations that connect to the student's concrete work.

■ Working It Out This New Way

1. Tell students that when using the distributive property, **the larger number is rewritten as a sum.** For example, in a two-digit number, it's written as a sum of the place values of the tens digit and the ones digit. [92 = 90 + 2.] The goal is to make the multiplication simpler.

Alge-Branch Point for Teachers

Have 3rd-graders continue to use base ten blocks to solve additional problems.

2. Return to the multiplication problem you posed at the start of class (for **3rd-graders,** 4 • 26; **for 4th,** 6 • 59; for **5th,** 9 • 87). Have your students solve the problem *using the distributive property.* Encourage 4th-grade students and higher to use their mental math skills to solve the problem. Outline the steps in the multiplication "trick," or distributive property:

Grade level	3rd	4th	5th
Record	4 • 26 =	6 • 59 =	9 • 87 =
Ask how to rewrite the two-digit number	4 • 26 = 4 • (20 + 6)	6 • 59 = 6 • (50 + 9)	9 • 87 = 9 • (80 + 7)
Multiply the value of each digit in the two-digit number by the single-digit number	= (4 • 20) + (4 • 6)	= (6 • 50) + (6 • 9)	= (9 • 80) + (9 • 7)
Calculate the partial products	= 80 + 24	= 300 + 54	= 720 + 63
Add, to calculate the product	= 104	= 354	= 783

■ Using Friendly Numbers

1. Tell students they don't always have to separate the number into a sum of tens and ones. Sometimes they can use other "friendly" numbers to do the calculations easily.

2. Explain that in the case of the problem 4 • 26 (the problem third-graders have been doing), 26 also equals 25 + 1. Using the distributive property, it could be written as 4 • 26 = 4 • (25 + 1). Write this on the board.

3. Ask students to help you describe the steps to solve the problem. Record the steps as they explain:

4 • 26 = 4 • (25 + 1)	*Rewrite 26 as 25 + 1*
= (4 • 25) + (4 • 1)	*Multiply each by the single-digit number*
= 100 + 4	*Calculate the partial products*
= 104	*Add, to calculate the product*

4. Give students another problem to solve in their journals. Ask them to record the steps to solve the problem *using the distributive property,* even if they're able to use mental math to solve it.

5. Ask a student to explain to the class how he used this new "friendly numbers" method to solve the problem. Check that everyone follows his reasoning.

■ Writing the Distributive Property Algebraically

1. Using an example students worked on earlier, demonstrate how to express what they did algebraically. Begin by recording equations, such as:

$$3 • 46 \qquad = 3 • (40 + 6)$$
$$46 + 46 + 46 = (3 • 40) + (3 • 6)$$
$$138 \qquad = 120 + 18$$
$$138 \qquad = 138$$

2. Remind them that algebra uses letters (variables) to represent numbers and to generalize rules.

3. For this problem, you'll use the letters *a*, *b*, and *c* as variables. Write *a* = 3, *b* = 40, and *c* = 6 on the board. Ask how to write 46 using these letters. [46 = *b* + *c*.]

4. Rewrite each equation and then substitute the variables into it as follows:

$$3 \bullet 46 \qquad = 3 \bullet (40 + 6)$$
$$a \bullet (b + c) \qquad = a \bullet (b + c)$$

5. Using the same method, distribute the a over the $(b + c)$ as follows:

$$a \bullet (b + c) \qquad = (a \bullet b) + (a \bullet c)$$

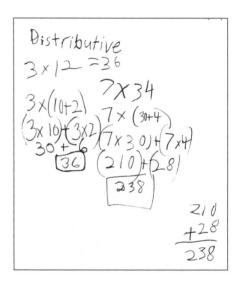

6. Ask students to explain this equation with a sentence. Listen and help students express it in their own words. During the discussion, be sure the class understands that the distributive property of multiplication over addition means that if their computations are accurate, they can get the same results by doing one of the following:

a. Multiply a number by a factor.

or

b. Break that number into a sum, multiply each addend by the factor, and add the two partial products.

7. Acknowledge that it may seem harder to express this property in words than it is to see it expressed in a mathematical way. **That's one of the strengths of algebra. Algebra is a way to generalize about ways that numbers work—using symbols to express ways of operating with numbers.** In this case, it's a way to express how Professor Arbegla's trick works for *all* numbers, not just for specific numbers.

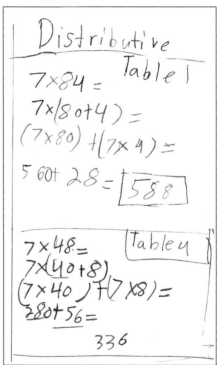

8. Record the distributive property on the **Algebra Tool Kit** chart, and have students add it to their journals, like this:

The Distributive Property: $a \bullet (b + c) = (a \bullet b) + (a \bullet c)$

9. For homework, have students apply the distributive property to several more problems. A few grade-appropriate examples follow. Have students take their journals home so they can refer to their tool kits when they work on the problems. Then collect the journals after Session 2.

Grade 3	Grade 4	Grade 5
$2 \bullet 53 = 2 \bullet (50 + 3) =$	$5 \bullet 47 = 5 \bullet (40 + 7) =$	$8 \bullet 56 = 8 \bullet (50 + 6) =$
$3 \bullet 65 =$	$6 \bullet 58 =$	$9 \bullet 73 =$
$4 \bullet 58 =$	$7 \bullet 49 =$	$7 \bullet 124 =$
$5 \bullet 47 =$	$8 \bullet 56 =$	$6 \bullet 453 =$

Session 2:
Revisiting the Distributive Property

1. Review the distributive property by having students present the solutions to at least two of their homework problems. Encourage them to include all the steps they used to solve each problem.

2. Ask for their ideas about how the distributive property was given its name. Encourage creative responses and build on any that touch on or connect to the explanation below. Let them know that the name comes from a description of what happens when we take a look at the algebraic expression of the property. Write the algebraic expression on the board.

$$a \cdot (b + c) = (a \cdot b) + (a \cdot c)$$

3. As students can see, on the left side of the equation the "*a*" outside the parentheses is the factor the other two numbers—*b* and *c*, inside the parentheses—are being multiplied by. On the right side of the equation, that factor "*a*" is **distributed** to the two terms within the parentheses.

4. Ask students if they think the distributive property will also work in the case of *subtraction*. As you take several responses, be sure to have students explain their thinking.

5. Have students return to a prior problem or use a problem where the ones digit of the two-digit number is greater than 5, such as 6 • 39. Record it on the board.

6. Review how to write the number 39 as the sum of two numbers, 30 + 9. Next, ask how to express 39 as the *difference* between two numbers. Ideally, students' answers will include 40 − 1. Substitute 40 − 1 for 39, as follows:

$$6 \cdot 39 = 6 \cdot (40 - 1)$$

7. Since the students know the distributive property over addition, ask them how they'd continue to solve this problem using the **distributive property over subtraction.** Provide time for them to solve the problem.

8. Have a student present the next steps to determine the solution, with support as needed from you and his classmates.

$$6 \cdot 39 = 6 \cdot (40 - 1)$$
$$= (6 \cdot 40) - (6 \cdot 1)$$
$$= 240 - 6$$
$$= 234$$

9. Show that this is the same answer you'd get if the problem was solved using addition.

$$6 \cdot 39 = 6 \cdot (30 + 9)$$
$$= (6 \cdot 30) + (6 \cdot 9)$$
$$= 180 + 54$$
$$= 234$$

10. Ask students which is easier for them when doing this problem—using $(30 + 9)$ or $(40 - 1)$. Say that when the digit in the ones place of the two-digit number is greater than 5, many people find it easier to use subtraction.

11. Provide your students with a few more problems to practice using the distributive property over subtraction:

Grade 3	Grade 4	Grade 5
$7 \cdot 19 = 7 \cdot (20 - 1)$	$7 \cdot 38 = 7 \cdot (40 - 2)$	$9 \cdot 49 = 9 \cdot (50 - 1)$
$6 \cdot 38 = 6 \cdot (40 - 2)$	$6 \cdot 59 = 6 \cdot (60 - 1)$	$8 \cdot 77 = 8 \cdot (80 - 3)$
$4 \cdot 29 = 4 \cdot (30 - 1)$	$9 \cdot 47 = 9 \cdot (50 - 3)$	$9 \cdot 88 = 9 \cdot (90 - 2)$

12. After most students have completed the problems, have volunteers present their solutions, being sure to explain how they used the distributive property for subtraction. Check that the rest of the class understands this property.

■ Generalizing with Algebra

1. Tell students that just as you wrote the distributive property for addition algebraically, you now want to write the distributive property for subtraction algebraically.

2. Go to the **Algebra Tool Kit** chart and read the property for addition with the students.

3. Have students work in pairs or independently to write the distributive property over subtraction algebraically.

4. After allowing time for students to work, have someone share her work. Have other students agree or disagree and *explain their reasoning.* The response should look like this:

$$a \cdot (b - c) = (a \cdot b) - (a \cdot c)$$

5. Record this property for subtraction on the **Algebra Tool Kit** chart and have students record it in *algebraic terms* in their journals.

6. Continue to give students several problems in which they can apply the distributive property. Have them decide which operation works best for each particular problem—addition or subtraction. Provide opportunities to share solutions for a problem using addition and subtraction. Over time, encourage them to use mental math skills to solve problems.

7. Collect the journals.

■ Going Further

The Distributive Property in Verse
a. Begin by brainstorming important words related to the distributive property. Record these on the board.

b. Share the poem on the next page with the class and suggest that they may want to write a poem expressing the distributive property.

Arbegla's Multiplication Trick

by Lincoln Bergman

Let me show you a little math trick
If you learn it, I think it'll stick
It's a neat way to multiply
Check it out and figure out why.

Suppose I decided to buy
Nine colorful kites to fly high
Each kite cost 94 cents
How much money will I have spent?

9 times 94 can be written and done
But here's how to have some more fun
Make 94 into 90 plus 4, that's fine,
Then multiply each of them by nine.

9 x 90's eight hundred and ten
9 x 4's thirty-six my friend
Add them up and what do you get?
Eight hundred forty-six, you bet!

With some practice do this in your head
You'll see how your mental math's sped
Make up problems when taking a bath
You'll soon wash off all fear of math!

Then multiplication will be smooth as a dream
You'll feel a new sense of math self-esteem!

And just to be fully informative
This property is called the distributive
Can you explain this long name?
Then you've a real claim to fame!

Letter #3: My Multiplication Discovery

Dear Mathematicians:

Wow, am I excited! I think I may have discovered an incredible trick that's helping me multiply more easily and quickly than I ever dreamed possible. I think I might even write a poem about it, I'm so inspired. Then I could **distribute** the trick through a clever poem!

But like any good scientist or mathematician, I need some other people to test my idea to be sure it works in all cases. I've tried it on some numbers, of course, and it always seems to work, but I really need you to confirm it. If it does work, maybe you and your teacher can help me figure out *why* it works.

Your teacher will explain—thanks, and may your luck multiply!

Multitudes of gratitude,

Professor Z. Arbegla

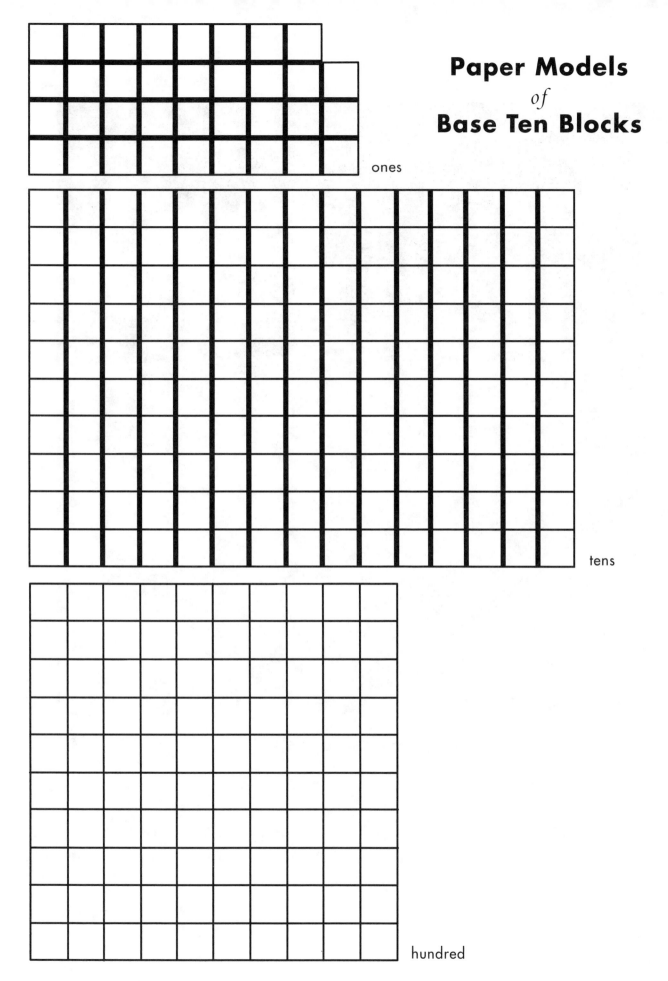

Paper Models
of
Base Ten Blocks

ones

tens

hundred

Overview

This activity opens with a story from Professor Arbegla's college class, in which a student asked the professor when and where algebra is used outside of school. That question made the professor think about home improvement projects—Professor Arbegla is currently remodeling her bathroom. As part of the work, the floor needs to be tiled. Algebra can help with that job! Your class has an opportunity to see algebra in action as they help the professor choose the exact number of tiles she'll need. In a new area problem, students determine the possible dimensions of rectangular rooms with a set area. They record all the possible (whole-number) lengths and widths, using grid paper. These geometric representations of area help illustrate the *commutative property of multiplication.* In subsequent activities, they continue to explore area and perimeter of rectangles.

In Session 2, students apply these dimensions to a real-world context. They use a standard unit of measure and, working as a class, measure and map the length and width of an enclosure to visualize its real size. They then measure the perimeter of the shape, and learn the distinctions between area and perimeter. Their understanding of these concepts is reinforced as they design animal enclosures for a preserve.

In Session 3, students solve area problems with variables to find an unknown in the equation—the length, width, or area. As students explore problems involving both area and perimeter, they have an opportunity to review how distinct these two functions are.

In this final activity, students build on their algebraic thinking and reasoning skills. They learn that, like addition, multiplication is commutative. Variables are used to write algebraic expressions and equations. As they use standard measurement, a square foot, they gain a better understanding of area and develop spatial sense. Students are also asked to think creatively as they apply mathematical information to a real-world problem.

If you've presented the GEMS guide Math on the Menu, your students have had an opportunity to design a restaurant floor plan, and will come to this problem with more understanding.

The commutative property of multiplication is described in "Background for the Teacher."

■ What You Need

For the class:
- ❏ the **Algebra Tool Kit** chart from Activity 5
- ❏ several wide-tipped colored markers
- ❏ 1 overhead transparency of one-centimeter grid paper (page 106, or you can purchase)
- ❏ 1 sheet of construction paper, 12 in. x 12 in.
- ❏ a ruler
- ❏ scissors
- ❏ 1 yardstick or tape measure
- ❏ string
- ❏ overhead pens
- ❏ an overhead projector

For each student:
- ❏ algebra journal from Activity 5
- ❏ approximately 4 sheets of one-centimeter grid paper (page 106 or purchased)
- ❏ 1 sheet of construction paper, 12 in. x 12 in.

■ Getting Ready

1. Make an overhead of the grid paper you copied (page 106) or purchased.

2. Set aside about four sheets of grid paper per student.

3. Using the same color of construction paper, cut a 12 in. x 12 in. piece for yourself and one for each student.

4. Decide where you'll take students—outdoors or in the school—to show them what an area 36 square feet looks like.

5. Gather papers, measuring tools, markers, and journals.

Session 1: Algebra in the Real World

1. While you're passing out the journals, tell your students that one of Professor Arbegla's college students recently asked her when and where he'd use algebra outside of math class. Before the professor answered, she asked her other students what they thought.

2. Give your class a chance to talk with a partner or in small groups about when and where they think algebra might be used outside of school.

3. Have students share their ideas with the class. Segue from their discussion to describing how Professor Arbegla's students responded.

4. Say that Professor Arbegla's students shared the following:

   • Celeste, a student majoring in chemistry, said she uses algebra in her classes to balance chemical equations

   • Matt reported that he uses algebra when he cooks—for example, when he has to increase or decrease quantities in a recipe

   • Amanda described how she used algebra to figure out the interest on her college loan

   • Tyrone, an architecture student, said he used algebra to find the area and perimeter of a volleyball court when he designed an addition for the athletics department

   • Mercedés said she uses algebra to calculate the batting averages of Major League players

5. Tell students that as her students talked, Professor Arbegla realized another use for algebra in the real world—home remodeling. The professor wants to retile her bathroom floor. She knows the dimensions of the floor and wants to pick out the tiles. She went to the

store to buy tiles and needed to know how many to buy. She didn't want to buy too many, as they were expensive, but she wanted to be sure to have enough for the entire floor.

6. Tell students the dimensions of Professor Arbegla's bathroom are six feet by seven feet. Ask them to determine how many one-foot-square tiles she'll need to cover the bathroom floor.

7. Some students may immediately know that 42 tiles are needed. Others may need to draw a picture. In any case, the important thing to ask is *how they came up with the number of tiles needed.*

In one fifth-grade class, students hadn't practiced determining area for a long time, and were also rusty on multiplication facts. Here's an example of how they arrived at the number of tiles needed for Professor Arbegla's bathroom floor.

Sample Dialogue:
How Many Tiles Does the Professor Need?

Teacher (TR): How can we determine the number of tiles Professor Arbegla needs?
Simone: Well, maybe we should draw a picture.

TR draws a rectangle on the board.

TR: Does this help?
Simone: Yeah and you should put the numbers on the sides.

TR records the length and width.

6

7

TR: How many tiles are needed to cover the floor? Please explain how you got that number.
Alejandro: I think 6 times 7.
TR: Why?
Alejandro: I don't know why...it's just 6 times 7.
Adriana: I think it's 26.
TR: How did you get that number?
Adriana: Because 7 times 2 is 14, and 6 times 2 is 12. And 14 plus 12 is 26.
TR: Where did those numbers come from?
Adriana: Well, there are two sides that are 7 and two sides that are 6.
TR: Does everyone follow Adriana's thinking?
Victoria: I still think it's 7 times 6.
Adriana: Oh, yeah, she's right.

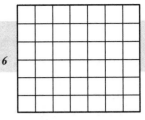

Adriana comes up to the board and draws lines to create 42 boxes within the rectangle.

6

7

Adriana [referring to the square units inside the rectangle]: All of these boxes are like the tiles, and there are 42.
TR: Do you all understand what Adriana has done?

Signs of agreement from majority of class.

TR: What are you figuring out when you do multiply 6 by 7? There's a term in math for this. Does anyone know what it is?

No responses.

TR: We're looking at the space **inside** the rectangle.
Jimmy: The area!
TR: How many people agree with Jimmy?

Agreement from most of the class.

TR: What is the area?
Cedric: It's the number of squares inside the rectangle.
TR: Is there anything special about those squares?
Cedric: I think they all have to be the same size or something like that.
TR: That's right. When we measure in square units, all the units have to be equal. Now, what did Adriana start us out with?
Eddie: I think that's something like the perimeter.
Adriana: That's right. What I did first was perimeter. You add up all the sides. When you multiply the sides, you get the area.
TR: How many agree with Adriana's explanation?

Agreement from most of the class.

TR: I agree with Adriana too. She said that to get the perimeter you add up all the sides. So in this case, the sides have a length of 6 feet and 7 feet. So the perimeter equals 6 + 7 + 6 + 7.

Records P = 6 + 7 + 6 + 7 on the board.

TR: How many feet does that equal?
Sharniece: 26 feet.
TR: Great. Now how can we write the area of this rectangle?
Maya: Well, like Adriana said, you multiply the sides, so it's 6 times 7.
TR: Maya, can you record the area on the board?

Maya comes to the board and records A = 6 • 7.

TR: So what is the area?
Maya: 42
TR: 42 what?
Maya: The area.
TR: Yes, it's the area. What does the 42 represent? Is that 42 feet, like the

perimeter, or something else? Can anyone help us?
Donte: Can I come to the board to explain?
TR: Of course!
Donte [pointing to the squares in the rectangle]: There are 42 squares, so the area is 42 squares.
TR: How many agree with Donte?

Signs of agreement. TR goes to the board by Donte.

TR: Let's look at those squares. What size is each one?
Donte: 1 foot by 1 foot.
TR: Exactly. So when we measure the area, we're measuring the square feet inside the rectangle or square or whatever shape it is. This rectangle has an area of 42 square feet.

Records A = 42 square feet on the board.

$$P = 6 + 7 + 6 + 7$$
$$P = 26 \text{ feet}$$
$$A = 6 \cdot 7$$
$$A = 42 \text{ square feet}$$

TR: OK, so now we know how many square-foot tiles Professor Arbegla needs. We also understand the difference between the area and perimeter of a rectangle. Let's write it using algebra.
TR: Let's let the length = l and the width = w. Talk to your neighbor and come up with a way to write an equation for the area and the perimeter.

Provides time for students to talk and work with a partner.

TR: Who'd like to come up and write one of the equations?
D'Nesha: I can do the area one.

D'Nesha goes to the board and writes A = $l \cdot w$.

TR: D'Nesha, can you tell us what that equation says in words?
D'Nesha: The area is equal to the length times the width.
TR: Great. What units is the area measured in?
Donte: Square units. Like square feet.
TR: Was anyone able to write the equation for the perimeter?
Adriana: I'll try.

Adriana goes to the board and writes P = $l + w + l + w$.

TR: Can someone tell us what that says in words? Adriana, call on someone.
Julissa: The perimeter is equal to the length plus the width plus the length plus the width.
TR: Do you all agree?

Signs of agreement.

TR: Did anyone write the perimeter a different way?
Alex: I wrote perimeter equals 2 times the length plus 2 times the width.

TR: Does this give us the same answer as what Julissa said? Why or why not?

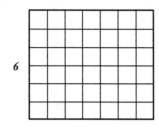

P = 6 + 7 + 6 + 7   P = *l* + *w* + *l* + *w*
P = 26 feet   P = 2 • *l* + 2 • *w*
A = 6 • 7   A = *l* • *w*
A = 42 square feet

Discussion continued until students had proven that the two equations are equivalent.

8. After discussing how to determine the number of tiles needed to cover the area of the bathroom floor, guide students to write the generalized formula for the area of a rectangle. Let *l* = the length; *w* = the width; and A = the area. Have students tell a partner how to write the equation. [A = *l* • *w*.] Next guide students to the generalized formula for the perimeter.

9. Ask a volunteer to tell you the equations, and record A = *l* • *w* and P = *l* + *w* + *l* + *w* on the **Algebra Tool Kit** chart. Have students record the equations in their journals.

10. Point out that once again, algebra helps us generalize. Whenever we want to find the area of a rectangle, we use the formula A = *l* • *w*. The perimeter can be determined using the formula P = *l* + *w* + *l* + *w*.

■ 36 Square Units: What's the Length? What's the Width?

1. Pose a new area problem for your students. Ask them how to write an equation in which the area of a room equals 36 square units. Record their answer on the board. [*l* • *w* = 36 or 36 = *l* • *w*.]

2. Since the area is known, ask students to find a length and a width for the room, **using whole numbers.** Have students talk with a partner, then ask the class for a length and a width. On the board, record the answer on a chart, as follows:

To provide an initial concrete experience for your students, you may want to have them make a 36-square unit using small (one-inch square), plastic tiles. Later students will draw 36-square foot rooms on grid paper.

$$\underline{l \bullet w \quad} = A$$
$$3 \bullet 12 \quad = 36$$

3. Check that students agree. Ask if there are other possible lengths and widths. Have students work in pairs to determine all the possible dimensions. Remind them to use only whole numbers.

4. Circulate as they work. When most students have generated the solutions, focus the class to discuss them.

5. Ask for another pair of dimension for which the area equals 36. Record the length and width on the chart on the board. If a student suggests 6 by 6, for example, then record as follows:

$$\underline{l \bullet w \quad} = A$$
$$3 \bullet 12 \quad = 36$$
$$6 \bullet 6 \quad = 36$$

6. Continue listing other solutions as students suggest them. A complete chart will include the following dimensions:

$$\underline{l \bullet w \quad} = A$$
$$3 \bullet 12 \quad = 36$$
$$6 \bullet 6 \quad = 36$$
$$12 \bullet 3 \quad = 36$$
$$4 \bullet 9 \quad = 36$$
$$2 \bullet 18 \quad = 36$$
$$9 \bullet 4 \quad = 36$$
$$18 \bullet 2 \quad = 36$$
$$1 \bullet 36 \quad = 36$$
$$36 \bullet 1 \quad = 36$$

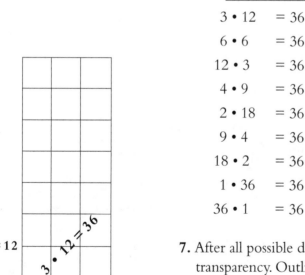

7. After all possible dimensions have been added, show the grid-paper transparency. Outline one square. Tell students that each square box represents one square foot.

8. Demonstrate how to record the room with the dimensions $l = 3$ and $w = 12$. Draw the rectangle on the transparency and label the lengths of the sides. Inside the rectangle, write the equation $3 \bullet 12 = 36$.

9. Tell students they'll record this rectangular room on a sheet of grid paper, label the dimensions, and write the area equation inside the rectangle. Distribute grid paper to each student and have them begin.

10. When students have completed this room, have them draw each of the remaining eight rooms on their grid papers.

11. Circulate as they work, assisting as necessary. Someone will probably notice that the dimensions can appear in pairs, such as *l* = 12 and *w* = 3, and *l* = 3 and *w* = 12. Have them look at the room formed with each pair of dimensions. Even though one is long and narrow and the other is wide and short, they have the same area!

12. Have students check that they have drawn all nine rooms. Ask if each room has a "partner" room—a room with the dimensions reversed. They'll probably notice that one room has both a length of 6 and a width of 6.

13. Take a minute to focus class attention on this rectangular 6 by 6 room. Does it have another geometric name? [Square.] Ask students how they can be certain it's a square. Agree that the following is true: a square is a special type of rectangle; all its sides are equal in length and all its angles measure 90 degrees.

14. Finally, have students look again at a pair of rooms with identical lengths but for opposite dimensions—such as *l* = 4 and *w* = 9, and *l* = 9 and *w* = 4. Ask, "Is the area of both these rectangles the same?" [Yes, both equal 36.] "Does it matter which order you multiply the numbers in?" [No.]

15. Check another pair of dimensions, such as 18 • 2 and 2 • 18. Again, students will see that it doesn't matter what order the numbers are multiplied in; the result is the same—36. In fact, the shapes of the rooms are *congruent* (the same size and same geometric shape). (By cutting them out and situating one on top of the other, students can prove this for themselves.)

16. Remind students of the commutative property of addition, referring to the **Algebra Tool Kit** chart. Tell students that, like the commutative property of addition, there's a *commutative property of multiplication.* If *a* and *b* represent numbers, then we can write the product of those numbers algebraically, as follows:

$a • b = b • a$

17. Record the commutative property of multiplication on the **Algebra Tool Kit** chart and have students record it in their journals. Then collect the journals.

For some students, it may be beneficial to have them actually cut out the two rooms with the same numbers in different dimensions and place them on top of one another to concretely see that the area is the same.

Depending on your students' prior knowledge, you may want them to focus on this square room. Since all sides are equal, the area can be expressed as "side times side"—which provides a square number. In a square, A = s • s, or s².

Challenge for fifth-graders:
For homework, ask older students to find four sets of dimensions that are not whole numbers for a room with an area of 36 square feet. Assign the homework to be done on loose paper and collect the students' journals, or have them work in their journals and collect them at the end of the activity.

Session 2:
The Many Shapes of 36 Square Feet

If your fifth-graders did the special challenge as homework in the previous session, start by reviewing the assignment. Ask what non-whole-number dimensions they came up with. Encourage them to share how they determined the dimensions.

1. After you've distributed the journals, tell the class you're going to give them a real-world reason to use the whole-number dimensions they generated in the last session. Say that six new animals have arrived at the local animal preserve. Each animal is of a different species, but each needs an enclosure with 36 square feet of room. The size and shape of each animal will determine the dimensions of its enclosure.

2. Define the unit of measure as one square foot. Show them your sheet of construction paper. Put a ruler along both sides to demonstrate that it is one foot wide by one foot long—one foot square.

3. Pass out the sheets of construction paper to the students. Let them know their paper is also one square foot. Have them record their names on the squares.

4. Tell students you want them to get a clear understanding of what 36 square feet looks like. Pick up the yardstick or tape measure and string, then lead the students outdoors or to a large room. Ask them to bring their construction-paper squares along.

You'll need a total of 22 squares to create these 9 by 4 dimensions. (Remember that the square in each corner accounts for two units of measure in the perimeter.) If you have fewer than 22 students, make the extra number of squares you'll need to total 22.

5. Model how to map out the actual size of an enclosure using one set of dimensions, such as nine feet long by four feet wide ($l = 9$ and $w = 4$).

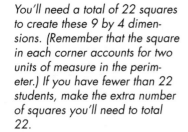

   a. Place your square foot piece of construction paper on the ground. Have another student place hers next to yours.

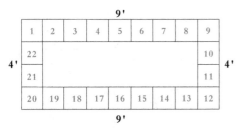

   b. Continue placing squares down in a row until there are nine, for length.

c. Create the width of the rectangle by adding squares perpendicular to the length, as follows:

d. Continue to build the dimensions of the 36-square foot enclosure so that the remaining length and width complete the perimeter of the rectangle.

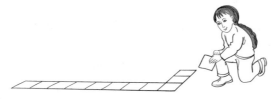

This gives students a concrete model of the size of an enclosure nine feet long by four feet wide.

6. Review what area is. Ask what perimeter is, and have a student point out the perimeter of the enclosure you just created.

7. Use the string to measure the full length of the perimeter; lay the string down around the perimeter of the rectangle, and cut off any excess.

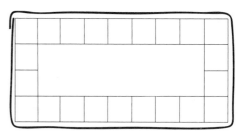

8. Have two students each take one end of the string and step away from each other to straighten it into one long length. Have them stretch it flat on the ground and use the tape measure or yardstick to measure the length. [P = 26 feet.] This clearly shows the perimeter as a length, or linear measurement, and concretely illustrates the difference between square and linear measurement.

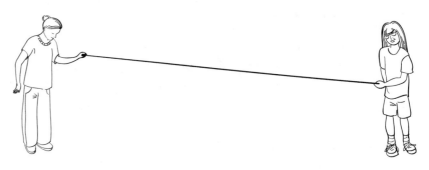

9. Remind the students about the six new animals at the animal preserve. Name the following animals: snake, mouse, lizard, coyote, rabbit, and turtle.

10. Ask which of these animal(s) might be suited to live in a habitat the size and shape of the area the class just created. Encourage rationale for students' choices.

11. Have students collect their construction-paper squares. Say they'll now create an area with dimensions as follows: *l* = 6 and *w* = 6. Follow the same procedure as before. Place squares in a line until there's one length, then complete the other three sides.

12. How does this square compare to the rectangle they just made? [Though they look very different and have different perimeters, they're equal in area. A = 36 square feet.]

13. Have students measure the perimeter using string, and compare the square's perimeter measurement with the perimeter of the rectangle. [P = 24 feet.] Which is longer? [The 9 by 4 rectangle has a greater perimeter.]

14. Be sure students know that even when two shapes have an equal area, they can look very different and have different perimeters.

15. Collect the squares and return to the classroom.

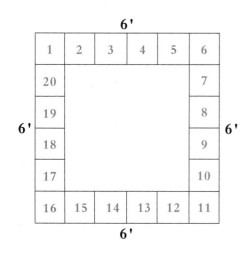

16. For homework, have students continue to explore the size and shape of each of the habitats with an area of 36 square feet:

> 1 by 36
> 2 by 18
> 3 by 12
> 4 by 9
> 6 by 6

Have them apply what they've learned: If they were the keepers in the animal preserve, what new animals would be best suited for each habitat shape, and why? Ask them to work in their journals or on another sheet of paper.

Session 3: More Variables

1. If you collected them at the end of the previous session, distribute the journals. Tell students that they can use algebra to solve problems related to area if they're given two numbers in the equation.

2. Pose the following problem as an example. A room was going to be re-carpeted. The owner knew the area was 84 square feet, and that the width of the room was seven feet (or six; see note). What's the length of the room?

> A = 84 square feet
>
> w = 7 feet
>
> l = ? feet

For third-graders, using 7 as the width and 84 as the area reinforces a multiplication fact, 7 • 12. However, you may want to ask fourth- and fifth-grade students to use 6 as the width, so they have to go beyond the multiplication tables.

3. Have students talk to a partner about how they would find the length. Ask a volunteer to explain his solution to the class. As necessary, guide students to the formula for area, A = l • w. Then substitute the known values into the equation: 84 = l • 7 (or 84 = l • 6).

4. Ask how they can solve for l, the length. Some students may say they know the multiplication fact, 7 • 12 = 84, so l = 12. Others may say that if you divide 84 by 7, the answer is 12—and therefore, l = 12. Remind them that division is the inverse operation of multiplication. Show the steps on the board as follows:

$$l \bullet 7 = 84$$
$$\frac{(l \bullet 7)}{7} = \frac{84}{7}$$
$$l = 12$$

For students in grades 4 and 5, you can provide a more challenging square, such as 121, and tell them that the l = w.

You might want to read the book, Spaghetti and Meatballs for All (see "Resources," page 130) before you do this activity. In that story, perimeter is explored in a humorous context.

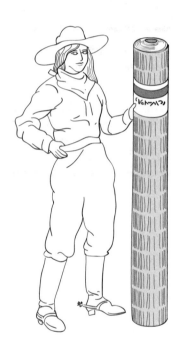

5. Pose another problem in which students can practice solving for an unknown in an area equation. Here's an example. If a room is 81 square feet and one length is 9 feet, what's the other length? This problem provides another opportunity to look at square numbers.

6. Pose additional ability-appropriate problems for students to work on in their journals. Have them design, in class or as homework, problems for their classmates to solve.

■ Perimeter Parameters

1. Review the formulas for area and perimeter. Record them so they're accessible to students.

2. Tell the class that Rancher Lynn wants to build a corral for her two cows. Taking into account the cost of fencing for the corral, Lynn determined that she could buy 80 feet of fencing. Lynn wants to give the cows as much grazing space as possible. What dimensions for the corral would create the largest area inside?

3. Have students work independently or with a partner to determine the best dimensions for the corral.

4. Circulate as they work. Ask clarifying questions to help students who may be having difficulties.

5. When most students have found a solution, have them present their solutions to the class. Be sure they explain how they determined the dimensions for the greatest area.

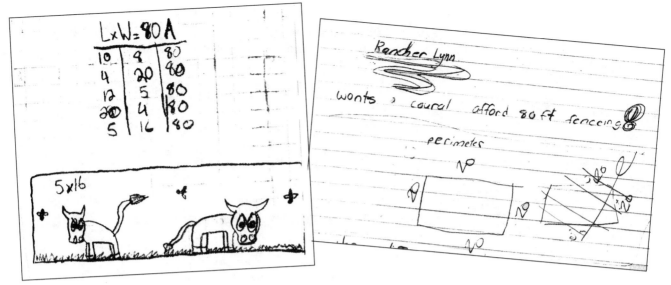

*Note the student's error of using 80
as the area, not the perimeter.*

■ The Dude Ranch (for Grades 4 and 5)

1. Pose another problem for your students, this time with an added twist
   to the problem. The owner of a dude ranch wanted to build a rectangular corral along the river on his property, so his horses would have
   access to water while in their corral. He purchased 90 feet of fencing.

2. Using the entire length of fencing the rancher has, ask students to
   create three possible rectangular corrals that border on the river. Have
   them draw a picture of each and label the *length* and *width* of each.

3. Circulate as they work. When all students have at least one corral,
   focus the class to share their solutions.

4. Ask a volunteer to come to the board and explain one of her solutions. As she explains, support her in drawing the corral and labeling
   it. Have the class check that all 90 feet of fencing are used.

5. Invite classmates to ask questions and make comments. Then ask for a
   different solution. Have another volunteer come to the board and
   present.

6. Continue to call on students until there's an assortment of rectangular
   corrals on the board.

7. Have student look at each corral. Which do they think would be best
   for the horses, and why? This should lead to a discussion of the area
   within the corral. If not, pose a guiding question, such as "Of all the
   corrals you've created, which one creates the greatest area?"

8. After comparing the areas visually and by using the formula for area, write on the board and ask students to think about the following:

   - Have we found the corral with the greatest area possible?

   - If so, how can you prove that?

   - If not, how can you find the corral with the greatest area?

9. Before taking responses, provide time for students to discuss this with partners to clarify and develop their explanations.

10. Circulate as they work. When most students seem ready, begin a discussion. Ask a volunteer to take a position on one of the questions posed. Invite him to go to the board to help explain his thinking.

11. Encourage other students to ask questions and check that they understand what's presented. Continue calling on students to present their thinking. Serve as the facilitator of the debate, building on students' ideas or asking strategic questions.

12. Depending on your students' abilities, use tables to record the lengths and widths and the resulting perimeters and areas. Have students explore how the length and width of the corral affects the area.

13. Continue posing problems that use variables. Include some that relate to area and perimeter as well as other grade-appropriate problems.

14. Collect the journals.

■ Special Going Further

Design a Bedroom Floor Plan

If your students have done the GEMS unit Math on the Menu, *they'll be familiar with the concept of a floor plan. If not, Activity 4, Session 1 in that guide would be a great follow-up for your class after they've designed their bedrooms.*

This activity provides an excellent opportunity for students to put what they've learned about area and perimeter to real-world use by designing a floor plan for the personal bedroom of their dreams. Be clear from the onset—especially for those students who may share a bedroom with other family members—that this is *not* for their *current* bedroom. This is a chance to imagine a new bedroom of their own, for which they get to decide the dimensions and how to arrange the furniture.

Though a room's dimensions may be equal (commutative property of multiplication), the way it's situated can make it look quite different! In addition, where the door is situated adds another variable to the room and its dimensions. As students work, they see and integrate these variables for themselves. Algebraic thinking, units of measure, scale, and dimensions (area and perimeter) come into play as they design their rooms from scratch. In a nice extension, they can be given a budget for shopping for items to furnish or embellish their rooms.

Since this is a multi-step problem, it's suggested that you assign it for homework as a "Problem of the Week," or have students work in class over the course of several work periods. It can also be started in class and completed at home.

The **Design a Bedroom Floor Plan** handout on pages 104–105 outlines the problem for your students. Before copying and distributing the handout, decide the due date for the assignment and fill it in. Provide 3–4 sheets of grid paper for each student. Decide if you'll have students do Part 3, the optional budget and shopping extension; if so, fill in the allotted budget amount on the second page of the handout.

Be sure to provide an opportunity for students to share their finished floor plans with their classmates, either in class presentations or by posting their floor plans around the room.

Using a scale grid is an opportunity for students to use a mathematical model to help them understand quantitative relationships. In this case, they're using proportional reasoning.

The Math Forum's Problems of the Week (POWs) are designed to provide creative, non-routine challenges for students in grades 3–12. For more on this resource, see page 130.

Design a Bedroom Floor Plan

DUE DATE: _____

Congratulations!

You have the opportunity to design the bedroom of your dreams. This bedroom is for you alone! No need to consult anyone about where you'll place your bed or other personal items.

This project is divided into three parts. There are directions in each part—be sure to follow them carefully.

As you design your room, you'll use **feet** as the standard unit of measure. That means you'll need measuring tools.

Enjoy designing your dream bedroom!

Part 1: Deciding Dimensions

Bedrooms are required to be *at least* 60 square feet, according to building codes. You're lucky! Your bedroom is larger—it has an area of 108 square feet. Think about the possible shapes you could have for your room. Keep in mind that the area inside the room must equal 108 square feet. Also, each of the sides of the room must be at least 6 feet in length.

1. Find at least **four** different possible sets of dimensions for your 108-square-foot room. You may use fractions or decimals. Round fractional parts to nearest *whole or half foot.*

2. Draw the four or more possible rooms on grid paper. Let each square unit equal one square foot—each unit is one foot in length and one foot in width.

3. Analyze the rooms and choose the one you want for your bedroom. **Save the drawings of all four rooms.**

4. On a clean sheet of grid paper, draw your favorite room to scale. Label the dimensions in feet.

5. Measure the width of any bedroom door in your home. Decide where the entrance to your dream room will be and mark the door on your grid.

Part 2: Adding the Basics

Your bedroom needs (at least) a bed and a dresser. Before deciding where you'd like to place them, you'll need to find out the actual measurements of each.

1. Measure the length and width of your current bed and the length and width of one or more dressers in your home.

2. Make a scale model of the bed and the dresser of your choice, using grid paper, and cut them out. Situate the bed and dresser in different locations on the scale drawing of the room you designed in Part 1, until you're happy with their locations.

3. When you've decided where you want your bed and dresser to go, record those locations in pencil on your grid-paper room.

Part 3: Final Additions (OPTIONAL)

Now for the fun part! You add the kinds of extra things that personalize your room.

A. To begin with, you may add **two additional pieces of furniture** to your room. Choose from such items as a desk, chair, nightstand, bookcase, or table. (A computer, television, or stereo is *not* considered a piece of furniture.)

  1. Measure the length and width of the two pieces of furniture you'd like to add. Make a scale model of each piece and check that the furniture fits in your room.

  2. Decide where you want to situate the two pieces and record the furniture on your grid-paper room.

B. You have a budget of $_____ to spend on additional items for your room.

  1. Look at advertisements in the newspaper (or visit various sources) and price the items you'd like. Be sure not to go over your budget!

  2. Make a list detailing how you spent your money.

  3. Decide where you'd like to situate your purchases in your room and draw them in.

  4. Gather **all the work you did** to complete your bedroom floor plan. Put your **final** bedroom floor plan on top. Staple or clip them together to take back to class.

1-cm Grid Paper

This section is designed to serve as a resource on the algebraic concepts presented in this unit, and can help you prepare to teach the activities. *It's not intended to be read aloud to students or copied for their use.* It may, however, have the answers to some of the questions your students pose. These questions can provide insight into different levels of prior knowledge, misconceptions, and anxieties (if any) about algebra.

What is Algebra?

Algebra is essentially generalized arithmetic. It expresses the universal validity of certain statements in mathematics—especially those about numbers—by the use of symbols.

Algebra as an Academic "Gatekeeper"

Algebra is one of the primary "gatekeepers" in tracking students along academic or non-academic pathways—a strategy that affects their entire lives. This has had a discriminatory impact on less advantaged students of all backgrounds. Given the realities of history, social structure, and the educational system, this negative impact has been strongest against historically underrepresented groups—including girls and young women. The film *Stand and Deliver* (Warner Bros., 1988) makes this point powerfully in regard to Chicano students in Los Angeles. The Algebra Project, EQUALS/FAMILY MATH, and many other educational, community, and parent organizations have grown over the past 25 years to help address these issues. See "Resources" on page 130 for more information. These groups are working to ensure equality of access to high-quality rigorous mathematics learning for all students—in other words, "algebra for all!"

The Vocabulary of Algebra

As with any subject, there are words essential to algebra that need definition. The following short list provides definitions for the terminology used in this unit.

Constant: Refers to a quantity that stays the same. The number of inches in a foot, for example, is a constant: there are always 12 inches in a foot.

Variable: Refers to a quantity that has a range of possible values. For example, the variable z can be used to represent the day's temperature. In San Francisco, for example, z has a range from above 0° and below 110° Fahrenheit.

Equation: An equation is a mathematical statement that two expressions are equal; for instance, $7 + 9 = 11 + 5$. Algebraic expressions involving one or more variables are equal for certain values of the variables. For example, $2x + 3 = 4x - 1$ expresses the equality of the statement for $x = 2$.

Function: A function describes a relationship in which the value of one variable depends on the value of another quantity. For example, the profit of a company in dollars depends (among other things) on the number of articles sold.

In algebra, the notation $f(x)$ means "function of x."

T-table: Tables are used in mathematics to organize data. A T-table (or T-chart) is a way to represent data about two variables. Organizing the table and looking for a pattern, or relationship, between the numbers can help to determine a function.

Real Numbers: The real number system is the set of all rational and irrational numbers. Real numbers can be represented on a number line, with irrational numbers being approximated.

The set of real number is infinite, and includes:

- *Natural or Counting Numbers:* positive whole numbers that start at the number one: 1, 2, 3….

- *Whole Numbers:* positive natural numbers, plus zero: 0, 1, 2, 3….

- *Integers:* negative and positive whole numbers: −2, −1, 0, 1, 2….

- *Rational Numbers* are numbers that can be expressed as a fraction or ratio, $\frac{a}{b}$ where $b \neq 0$. For example, $\frac{2}{1}$ (which equals 2); $\frac{3}{4}$; $-\frac{5}{15}$; .18 (fractional equivalent is $\frac{18}{100}$); etc.

- *Irrational Numbers:* numbers that cannot be represented by a fraction or ratio that have a terminating decimal. For example, pi ($\frac{22}{7}$ or 3.14….); square root of 2; −2.236…. These numbers can be approximated and situated on the number line.

This Venn diagram organizes the real number system:

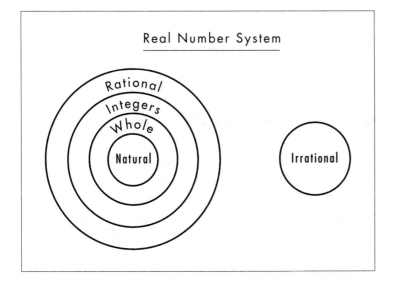

The Vocabulary of Computation

For each of the four arithmetic operations, the related equation is composed of parts with specific names, depending on their location in the equation. The following equations review the vocabulary of the various **parts of equations** for the four operations.

Addition Equation: Addend + Addend = Sum

Addend: any number being added to a number(s) to produce a sum.
Sum: the total; the result of adding at least two numbers together.

Subtraction Equation: Minuend − Subtrahend = Difference

Minuend: the number you subtract from.
Subtrahend: the number being subtracted.
Difference: the amount remaining after a quantity is subtracted from another quantity.

The terminology for subtraction isn't generally used in mathematics, and isn't important to emphasize.

Multiplication Equation: Factor • Factor = Product

Factor: a whole number that can be multiplied by another whole number to arrive at a product.
Product: the result when two whole numbers are multiplied together.

Division Equation: Dividend ÷ Divisor = Quotient

Dividend: a number that is divided by another number.
Divisor: the number by which another number is divided.
Quotient: the result of dividing one number by another number.

The Vocabulary of Plane Geometry

Plane geometry is the study of shapes and figures in two dimensions—
the plane. Plane figures have only length and width.

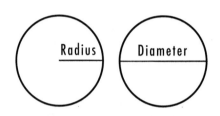

Circle is a plane figure bounded by a single curved line, every point of
which is equally distant from the point at the center of the figure. The
radius is the distance from the point in the center of the circle to any
point along the curved line. The *diameter* is a line segment passing
through the center of the circle from one point on the curved line to
another point on the opposite side.

Polygon is any closed plane figure with three or more straight sides.
"Poly" means many, and "gon" means sides, so the word polygon liter-
ally means "many sides."

Triangle is a plane figure with three sides and three angles. *The sum of all
three angles in any triangle equals 180°.*

> *Equilateral triangle* has all three sides equal in length and all three
> angles equal in degrees (60°).

> *Isosceles triangle* has two sides of equal length and two angles of
> equal degrees.

Right triangle has one 90° angle.

Acute triangle has all acute angles (less than 90°).

Obtuse triangle has one obtuse angle (more than 90° and less than 180°).

Scalene triangle has no sides of equal length.

Quadrilateral is any closed figure with four straight sides and four angles. *The sum of the angles in any quadrilateral equals 360°.* Quadrilaterals include the following.

Parallelogram has opposite sides that are parallel and equal in length.

Rhombus is a parallelogram with equal sides and opposite angles that are equal in measure.

Square is a quadrilateral with all sides of equal length and all angles of 90°. A square is also a rhombus.

Rectangle has all 90° angles, and opposite sides that are parallel and equal in length.

Trapezoid is a quadrilateral with only one pair of parallel lines.

Additional Polygons

A ***regular polygon*** has all sides and all angles equal. Polygons that have more than four sides are often portrayed as regular polygons. However, polygons don't have to have equal sides and equal angles. Polygons can have any number of sides, and are *named according to the number of sides they have.*

Pentagon: a polygon with **five** sides.

Hexagon: a polygon with **six** sides.

Heptagon (or septagon): a polygon with **seven** sides.

Octagon: a polygon with **eight** sides.

Nonegon: a polygon with **nine** sides.

Decagon: a polygon with **ten** sides.

Undecagon (or hendecagon): a polygon with **eleven** sides.

Dodecagon: a polygon with **twelve** sides.

"*N-gon*" is a polygon with *n* number of sides.

Properties of Numbers and Operations

The following properties are basic laws of operations for our number system. Primary students first encounter these laws informally and gain an intuitive understanding of them through concrete examples. In this guide, students have the opportunity to "name" and define these laws to develop a deeper understanding of them.

Identity Element

An identity element is a number that combines with any other number *in any order* without changing the original number. There are two identity elements, one for addition and one for multiplication.

The identity element for addition is 0.
  Example: $7 + 0 = 7$ and $0 + 7 = 7$
  This is expressed algebraically as:
    $a + 0 = a$ and $0 + a = a$ in all cases, *a* is a real number

The identity element for multiplication is 1.

Example: $7 \cdot 1 = 7$ and $1 \cdot 7 = 7$

This is expressed algebraically as:

$a \cdot 1 = a$ and $1 \cdot a = a$ in all cases, a is a real number

IMPORTANT NOTE: There are **no** identity elements for subtraction and division.

Inverse Operation

An inverse operation is an operation that is opposite from the original operation. *The inverse of addition is subtraction* and *the inverse of subtraction is addition.* In an equation or expression, given a quantity being added (or subtracted), the inverse operation is **subtracting (or adding) that same quantity.** The computational result of applying the inverse operation for addition or subtraction is zero.

For example, inverse operations are used to solve for n:

$$n + 3 = 9 \qquad\qquad n - 3 = 9$$
$$n + 3 - 3 = 9 - 3 \qquad\qquad n - 3 + 3 = 9 + 3$$
$$n + 0 = 6 \qquad\qquad n - 0 = 12$$
$$n = 6 \qquad\qquad n = 12$$

Similarly, *the inverse of multiplication is division* and *the inverse of division is multiplication.* In an equation or expression, given a quantity being multiplied (or divided), the inverse operation is **dividing by (or multiplying by) that same quantity.** The computational result of applying the inverse operation for multiplication or division is one.

Here again, for example, inverse operations are used to solve for n:

$$5 \cdot n = 40 \qquad\qquad n \div 5 = 40$$
$$(5 \cdot n) \div 5 = 40 \div 5 \qquad\qquad or \ \ \frac{n}{5} = 40$$
$$or \ \ \frac{(5 \cdot n)}{5} = \frac{40}{5} \qquad\qquad \left(\frac{n}{5}\right) \cdot 5 = 40 \cdot 5$$
$$\frac{5}{5} \cdot n = 8 \qquad\qquad \frac{5}{5} \cdot n = 200$$
$$1 \cdot n = 8 \qquad\qquad 1 \cdot n = 200$$
$$n = 8 \qquad\qquad n = 200$$

> In equations, if you perform an operation to one side of the equation, the same operation must be performed to the other side, to maintain the equality.

The Commutative Property

There is a commutative property for addition and multiplication only:

1. Commutative Property of Addition

This property states that the sum of the addends remains the same **regardless of the order in which they are added,** for all addends that are real numbers.

    Example:
        $15 + 23 = 23 + 15$
        $38 = 38$
    This is expressed algebraically as:
        $a + b = b + a$ in all cases, a and b are real numbers

A concrete example of commutivity is the distance one travels, or "commutes," from home to school or work (H ➜ S) and then from school or work to home (S ➜ H), assuming, of course, the same route is used. (H ➜ S = S ➜ H)

2. Commutative Property of Multiplication

This property states that for all factors that are real numbers, the product of the factors remains the same regardless of the order in which they are multiplied.

    Example:
        $7 \cdot 9 = 9 \cdot 7$
        $63 = 63$

    This is expressed algebraically as:
        $a \cdot b = b \cdot a$ in all cases, a and b are real numbers

IMPORTANT NOTE: The ONLY way to apply this property to **subtraction** or **division** is to *change the equation* and rewrite it as an addition or multiplication equation. For example:

    The following **subtraction** expression:
        $23 - 15$
    If rewritten as:
        $23 + (-15)$

could be transformed into the following commutative addition equation:

$$23 + (-15) = (-15) + 23$$

This is expressed algebraically as:

Given $a - b$

then $a + (-b) = (-b) + a$ in all cases, a and b are real numbers

Similarly, for **division,** the following expression:

$$25 \div 5$$

If rewritten as:

$$25 \cdot \tfrac{1}{5}$$

could be transformed into the following commutative multiplication equation:

$$25 \cdot \left(\tfrac{1}{5}\right) = \left(\tfrac{1}{5}\right) \cdot 25$$

Algebraically:

Given $a \div b$

then $a \cdot \left(\tfrac{1}{b}\right) = \left(\tfrac{1}{b}\right) \cdot a$ in all cases, a and b are real numbers and $b \neq 0$

Distributive Property

Multiplication is distributive over addition and subtraction. This property states that the product of a **number** and the **sum** (or **difference**) of two numbers is equal to the sum (or difference) of the products of the two numbers.

Example:

By the definition of multiplication, the expression $7 \cdot 28$ represents $28 + 28 + 28 + 28 + 28 + 28 + 28$.

Adding these numbers, the sum is equal to 196.

Since the number 28 can be expressed as either a sum of or difference between two numbers, as follows,

$28 = 20 + 8$ and $28 = 30 - 2$,

then either expression can be substituted into the expression $7 \cdot 28$ to find the value.

In this unit, the distributive property is used concretely to solve multiplication problems. The use of an area model and base ten blocks help illuminate this property.

By substituting 20 + 8 for 28, the distributive law can be applied and used to determine the value of the expression, as follows:

$$7 \cdot 28 = 7 \cdot (20 + 8)$$
$$= (7 \cdot 20) + (7 \cdot 8)$$
$$= 140 + 56$$
$$= 196$$

Similarly, expressing 28 as a difference, and substituting it in the expression, the distributive law can be applied as follows:

$$7 \cdot 28 = 7 \cdot (30 - 2)$$
$$= (7 \cdot 30) - (7 \cdot 2)$$
$$= 210 - 14$$
$$= 196$$

There are many other ways to rewrite the 28. For example, since 25 is an easy number to multiply, 28 can be expressed as 25 + 3.

Substituting 25 + 3, the distributive law can be applied again, as follows:

$$7 \cdot 28 = 7 \cdot (25 + 3)$$
$$= (7 \cdot 25) + (7 \cdot 3)$$
$$= 175 + 21$$
$$= 196$$

In the problems they encounter in the unit, students have an opportunity to see that the result remains the same regardless of the order of operations. If you do the addition or subtraction first and then multiply the result, your answer will be the same as if you multiply each addend or the minuend and subtrahend first, then do the computation. **In all these cases the result is the same.**

The distributive property is expressed algebraically as:

$a \cdot (b + c) = (a \cdot b) + (a \cdot c)$ in all cases, a, b, and c are real numbers
$a \cdot (b - c) = (a \cdot b) - (a \cdot c)$ in all cases, a, b, and c are real numbers

Using the distributive property can help make multiplication problems easier to solve mentally. This application of the property in grades 3–5 helps lay the foundation for later years, when students will encounter and use the property more abstractly.

How Did Algebra Get Its Name?

In the time of the Byzantine Empire lived a famous scholar named Mohammed ibn-Musa al-Khwarizmi. Al-Khwarizmi and his colleagues were scholars at the House of Wisdom, in Baghdad, and studied and wrote on algebra, geometry, and astronomy. The title of his treatise *Hisab al-jabr w'al-muqabala* in 825 C.E. gave us the word *"al-jabr"*—**algebra,** meaning "reduction" or "restoration"—and referred at that time to the process of removing negative terms from an equation. Over time, algebra came to refer to the study of equations. The term *"al-muqabala"* means **"balancing,"** the process of reducing positive terms of the same power when they occur on both sides of an equation.

In addition to his work in the area of algebra, al-Khwarizmi also wrote a book on arithmetic that became known in the 12th century by its Latin translation—*Algoritmi de numero Indorum*. (The word **algorithm,** meaning a rule for computation, is derived from *"algoritmi,"* the Latin form of al-Khwarizmi's name.) This book introduced Western Europe to Hindu-Arabic numerals and computations that use them—which is our arithmetic today! (Interestingly, in the 15th and 16th centuries, the Italians distinguished algebra from arithmetic by calling algebra the *greater art* and arithmetic the *lesser art*.)

In Spain, the Moors introduced a more common meaning of the Arabic term *"al-jabr."* The word *"algebrista"* evolved to mean a person who re-set, or "restored," broken bones. The word found its way to Italy in the 16th century, and in Europe the word "algebra" expanded to mean the art of bone setting.

In the 18th century, Swiss mathematician Leonhard Euler's book, *Instructions in Algebra,* became the model of algebra texts and firmly established the name for both Arabic and European scholars. (He also popularized the symbol for pi, invented by English mathematician William Jones.) You may want to encourage your students to do some research on the story of algebra, as it affirms the rich tapestry of multicultural and multinational contributions to mathematics and science. There were many other world mathematical influences that went into creating what we call algebra today! ∎

TEACHER'S OUTLINE

ACTIVITY 1: THE FABULOUS FUNCTION MACHINE

Session 1: Professor Arbegla's Function Machine

■ Getting Ready
1. Make Our Ideas and Our Questions charts; draw Fabulous Function Machine.
2. Make or obtain journals. Gather markers, tape, and charts.

■ Introducing Algebra and the Function Machine
1. Distribute journals.
2. Discuss algebra.
3. Have students write what they know and their questions about algebra.
4. Record on charts.
5. Introduce Professor Arbegla, the Fabulous Function Machine, and a T-table.
6. Ask for number from 1 through 50. If even, "0" is output; if odd, output is "1." Record on T-table.
7. Introduce one-step addition function.
8. Take numbers, process, have students discuss and record in journals.

Session 2: Decoding Functions

■ Getting Ready
1. Make Tool Kit chart. Decide on functions to use.
2. Gather markers, tape, and Tool Kit chart.

■ Geometric Shapes and the Function Machine
1. Introduce geometry and polygons.
2. Transform suggested polygon by "add 1" rule.
3. After students grasp rule, show how to write it algebraically.
4. Continue with other challenging functions, encouraging discourse and assisting students in writing down functions as algebraic expressions and describing in words.
5. Introduce Tool Kit chart and have students draw in journals.

ACTIVITY 2: MALFUNCTIONS IN THE FUNCTION MACHINE

Session 1: What Goes In Comes Out (Letter #1)

■ Getting Ready
1. Copy Letter #1, have it signed, and make overhead.
2. Have markers available.

■ Letter #1 from Professor Arbegla

1. Read letter aloud and show on overhead.
2. Focus on T-table. Have students discuss, then draft letter to Arbegla to explain what is going on with machine.
3. Discuss student ideas and record expressions on Tool Kit.
4. Explain $n + 0 = n$ as identity element for addition, and $n \cdot 1 = n$ as identity element for multiplication. Add these to Tool Kit and have students add to journals.
5. Have students write more polished letter.

Session 2: Oh, No, Not Again! (Letter #2)

■ Getting Ready

1. Copy Letter #2, have it signed, make copies and overhead.
2. Decide how much time for work at home.

■ Another Letter! Oh My!

1. Read letter aloud and show on overhead.
2. Distribute journals and Letter #2. Outline expectations for work at home and state due date.

■ Discussing the Letter

1. Have students share ideas about machine with partner or group.
2. Ask student to come to board and create T-table.
3. Ask students to share ideas about what machine is doing.
4. As needed, scaffold steps to algebraic expression.
5. Challenge them to find other solutions. Provide $7 + n - n$ expression.

ACTIVITY 3: THE MORPH MACHINE

Session 1: A New Math Machine

■ Getting Ready

1. Draw Morph Machine.
2. Gather tape and Morph Machine chart.

■ Introducing the Morph Machine

1. Introduce Morph Machine.
2. Discuss meaning of "morph" and explain/demonstrate machine.
3. Ask students to provide some morphed numbers to restore.
4. Continue "reprogramming" machine, processing, recording, and having students discuss and record in journals.

Session 2: What's the Magic in the Morph?

■ A New Way to Record
1. Review Morph Machine and how it works.
2. Write "*n*" and "Morphed *n*" on board.
3. Give calculations and have students morph numbers, as specified in guide.
4. Guide students to work backward from last step, or start at first.
5. Ask for different morphed number and record. Challenge students to determine original number. Circulate as they work, then discuss.
6. Guide students in backward problem-solving strategy; introduce inverse operations.
7. The machine's "magic" is ALGEBRA and they're using it!
8. For older or experienced students, show algebraic way to write/simplify each step.

■ More Morphing and Restoring
1. "Reprogram" machine to do another series of operations. Continue as before.
2. Ask what students want to add to Tool Kit. Add same information to journals.
3. Assign creating a process to morph and restore numbers as homework.

ACTIVITY 4: PROFESSOR LABARGE'S SCALES

■ Getting Ready
If using beans, fill each container with $\frac{1}{4}$ cup. Set aside two containers per student pair.

Session 1: Professor LaBarge's Scale

1. Introduce Professor LaBarge and her scale. Draw a scale on board. Call attention to "equal sign."
2. Discuss how a scale works.
3. Ask how scale could be balanced. (Be sure they know weights on each side must be equal.)
4. Challenge students with first scale problem.
5. Circulate. As students get one solution, encourage them to look for others.
6. Discuss. Point out that algebra allows us to generalize relationships between these weights.
7. Introduce next scale challenge. After solution, record the equation the scale represents.
8. Introduce commutative property of addition via work and discourse on next scale problem.
9. Have students explain property; add to Tool Kit and journals.

Session 2: More Balancing Acts

1. Challenge students with Problem #1 and discuss possible solutions.
2. Give students Problem #2 and discuss, using both equations and in words.
3. Present Problem #3. Encourage more than one solution.

ACTIVITY 5: THE DISTRIBUTIVE PROPERTY— PROFESSOR ARBEGLA'S GREAT MULTIPLICATION DISCOVERY

■ Getting Ready

1. Copy Letter #3 and have it signed.
2. Gather base ten blocks or make paper models.
3. Read background information on distributive property.
4. Gather letter and markers.

Session 1: The Professor's Multiplication Discovery

1. Introduce Arbegla's new multiplication "trick" and read aloud her letter.
2. Ask students to mentally calculate product of a single digit and a two-digit number.
3. Explain how to use blocks to demonstrate distributive property over addition.
4. Continue providing problems, recording, and eliciting student explanations.
5. Point out that larger number is rewritten as a sum and that "friendly" numbers can help.
6. Demonstrate and explain how to express this algebraically.
7. Elicit student participation, discussion, explanations.
8. Record property on Tool Kit, and in journals, as: $a \cdot (b + c) = (a \cdot b) + (a \cdot c)$
9. Assign several more problems as homework.

Session 2: Revisiting the Distributive Property

1. Review by students presenting solutions to at least two of homework problems.
2. Ask for ideas about then explain how distributive property was given its name.
3. Ask students if this will also work with subtraction.
4. Discuss and provide examples to show how to use distributive property over subtraction.
5. Have students write this property algebraically.
6. Have students share work and explain reasoning.
7. Record property on Tool Kit, and in journals, as: $a \cdot (b - c) = (a \cdot b) - (a \cdot c)$
8. Continue giving more problems; encourage mental math skills.

ACTIVITY 6: ALGEBRA IN ACTION

■ Getting Ready

1. Make overhead of grid paper, and gather three sheets of grid paper per student.
2. Cut a construction paper square for yourself and each student.
3. Gather papers, measuring tools, markers, and journals.

Session 1: Algebra in the Real World

1. Referring to Professor Arbeglas' students, discuss real-world uses of algebra.
2. Pose bathroom tile problem and guide students to formula for area of a rectangle.
3. Pose the first 36 square unit problem; students determine all possible dimensions.
4. Demonstrate how to record dimensions on grid paper and have students draw all nine rooms.
5. Have students look at pairs of rooms with similar dimensions.
6. Introduce commutative property of multiplication, expressed algebraically, as: $a \cdot b = b \cdot a$.
7. Record on Tool Kit and in journals.

Session 2: The Many Shapes of 36 Square Feet

1. Introduce six animals problem.
2. Model how to map out enclosure as described in guide.
3. Review area and perimeter. Use string to measure full length of perimeter.
4. Discuss results and possible spaces for different animals.
5. Have students map out a new enclosure with different dimensions and measure perimeter with string.
6. Point out that even when two shapes have equal area, they can look different and have different perimeters.

Session 3: More Variables

1. Pose carpet example and discuss as in guide.
2. Pose another problem to practice solving for an unknown in an area equation.
3. Pose more problems to work on in journals. Have them design problems for classmates.

■ Perimeter Parameters
1. Review and record formulas for area and perimeter.
2. Pose Rancher Lynn corral problem. Circulate as students work.
3. When most have found a solution, have them present and explain.

■ The Dude Ranch (for Grades 4 and 5)
1. Pose dude ranch corral problem.
2. When all have at least one corral, focus class to share solutions.
3. Ask volunteer to come to board and explain.
4. Ask for a different solution. Continue until there's an assortment of corrals.
5. Have they found corral with greatest area?
6. Discuss. Ask students to explain thinking.
7. Record lengths, widths, perimeters, areas.
8. Continue posing problems with variables.

ENCOURAGE ALGEBRAIC REASONING!

ASSESSMENT SUGGESTIONS

Anticipated Student Outcomes

1. Students demonstrate understanding of **variables** and are able to use symbols and letters to represent variables. Students are able to identify **algebraic expressions and functions** and express them in words and algebraically. They gain familiarity with **T-tables** and are able to use them as a tool to represent and determine algebraic expressions and functions.

2. Students develop an understanding of the following **properties of Numbers and Operations:**

- Identity Element for Addition and Multiplication

- Commutative Property of Addition and Multiplication

- Distributive Property over Addition and Subtraction

3. Students demonstrate increased understanding of **equations** and are able to **solve an equation** that has an unknown.

4. Students are able to **apply their understanding of algebra and their ability to reason algebraically** to known formulas, such as area.

5. Students are able to explain and apply the use of **inverse operations** as a problem solving strategy. They deepen their understanding of **computational strategies** and hone their **mental math skills.**

Embedded Assessment Activities

As you teach and facilitate the activities in this guide, you may need to adjust the content to fit the skills and abilities of your students. This may mean taking a step back, or adjusting the numbers students work with, or providing more challenging problems. As the teacher, you are constantly assessing as you implement the curriculum.

Algebra Journal. The algebra journal in which students document their classwork and homework throughout the unit, and to which they add "tools" for their own Algebra Tool Kits, provides an excellent, holistic, and ongoing assessment instrument. Throughout the unit, students express in words and numbers the algebraic and computational

Get Connected – Free!

Get the *GEMS Network News*,

our free educational newsletter filled with...

- **updates** on GEMS activities and publications
- **suggestions** from GEMS enthusiasts around the country
- **strategies** to help you and your students succeed
- **information** about workshops and leadership training
- **announcements** of new publications and resources

Be part of a growing national network of people who are committed to activity-based math and science education. Stay connected with the **GEMS** Network News. *If you don't already receive the* **Network News,** *simply return the attached postage-paid card.*

For more information about GEMS call (510) 642-7771, or write to us at GEMS, Lawrence Hall of Science, University of California, Berkeley, CA 94720-5200, or gems@uclink.berkeley.edu.

Please visit our web site at www.lhsgems.org.

GEMS activities are effective and easy to use. They engage students in cooperative, hands-on, minds-on math and science explorations, while introducing key principles and concepts.

More than 70 GEMS Teacher's Guides and Handbooks have been developed at the Lawrence Hall of Science — the public science center at the University of California at Berkeley — and tested in thousands of classrooms nationwide. There are many more to come — along with local GEMS Workshops and GEMS Centers and Network Sites springing up across the nation to provide support, training, and resources for you and your colleagues!

Yes!

Sign me up for a free subscription to the

GEMS Network News

filled with ideas, information, and strategies that lead to Great Explorations in Math and Science!

Name_____

Address_____

City_____ State_____ Zip_____

How did you find out about GEMS? (Check all that apply.)

❏ word of mouth ❏ conference ❏ ad ❏ workshop ❏ other: _____

❏ In addition to the *GEMS Network News,* please send me a free catalog of GEMS materials.

❏ Also, sign me up for the online edition of the *GEMS Network News* at this e-mail address:_____

GEMS
Lawrence Hall of Science
University of California
Berkeley, CA 94720-5200
(510) 642-7771

Ideas ◄
Suggestions ◄
Resources ◄

that lead to Great Explorations
in Math and Science!

**Sign up
now for a
free subscription
to the *GEMS*
Network News!**

101 LAWRENCE HALL OF SCIENCE # 5200

1-61571-25775-62-X

Get Connected!

www.lhsgems.o

concepts they acquire, and transfer algebraic tools with which to work on assigned problems and extension activities. The "Special Going Furthers" (pages 53 and 102), and "Additional Assessment Activity" (pages 128–129) can be added to or written in the students' journals, providing a clear measure of how well they've understood and implemented the elements of algebraic reasoning introduced in the unit.

Classroom Discourse. Throughout the unit, you are encouraged to initiate and guide class discussions that elaborate on students' questions and predictions. These discussions provide ideal assessment opportunities, as you gauge students' abilities and skill acquisition relative to the content you've taught, as well as their ability to articulate algebraic concepts and reasoning.

There are also specific, built-in assessments in each activity, including the following strategic assessments and the anticipated outcomes they address.

Letters to Professor Arbegla. (Activity 2, Sessions 1 and 2) Students first draft letters to Professor Arbegla explaining the first "malfunction" in the Function Machine. After a class discussion, they have a chance to edit and polish a final letter to the professor explaining what the function machine is doing. This provides both a pre- and post- assessment. Students respond to a second letter from the professor, about another "malfunction" in the machine. After a classroom discussion, students find additional solutions to this malfunction. (Addresses outcomes 1, 2, 3)

Decode the Morphed Number. (Activity 3, Sessions 1 and 2) Students are challenged to determine the original number entered into the Morph Machine by working back from the transformed, or "morphed," number that comes out. (Outcomes 1, 3, 5)

More Balancing Acts. (Activity 4, Session 2) Using a scale as a model for an equation, students solve equations with variables. (Outcomes 1, 2, 3, 5)

Apply the Distributive Property. (Activity 5, Sessions 1 and 2) Students solve problems, in class and as homework, involving the distributive property. (Outcomes 1, 2, 3, 5)

Habitats for Animals. (Activity 6, Sessions 1 and 2) For homework, students determine the various enclosures that can be created given an area of 36 square feet, and then decide what animal is best suited to live in a habitat of each resulting shape in an animal preserve. (Outcomes 1–5)

More Variables. (Activity 6, Session 3) Students solve problems related to area that involve equations with variables. (Outcomes 1–5)

Perimeter Parameters. (Activity 6, Session 3) Given the perimeter, students determine lengths and widths of a corral to decide which dimensions will provide the greatest area. (Outcomes 1–5)

Additional Assessment Ideas

Design a Bedroom Floor Plan. See the "Special Going Further" on page 102.

Balancing Mobiles. Use mobiles, instead of scales, for students to "balance" equations. Create mobiles appropriate to your students' skills and abilities. Here are several examples to get you started:

Mobile Level 1: $\triangle$ + $\square$ = $\diamondsuit$

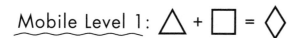

- *Provide values for*
  $\diamondsuit$ *and* $\triangle$. *Find* $\square$.

- *Use numbers appropriate to your students' abilities.*

- *Only identical shapes have equal weight.*

- *Each shape has a weight.*

Mobile Level 2: $\triangle$ + $\odot$ = $\diamondsuit$ + $\square$

- *Provide values for 3 of 4 variables or shapes.*

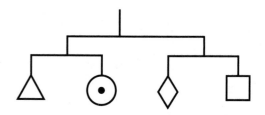

Mobile Level 3: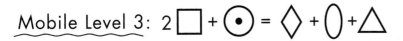

• *Provide values for the following shapes*

• *Challenge students to find more than one solution.*

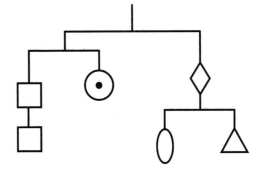

What Comes Next? To help provide an assessment of your students' overall comprehension and skills acquisition, the following activity provides an opportunity for them to apply what they've learned throughout the unit.

What Comes Next?
An Additional Assessment Activity for the Whole Unit

Name: _____

Using tiles, **Alex has started to build a pattern.** He'd only built the first two parts of his pattern when he was interrupted. Here's what Alex has built so far:

1. **Look carefully at the pattern.** What do you think Alex would build next? (You can use tiles or another math tool to help you.)

2. **Record your idea for the next part of the pattern.**

3. **Continue this pattern.** Record the next two parts.

4. **Explain your pattern in words.** How does the pattern grow? How do you know how to build what comes next?

_____

_____

_____

_____

_____

5. Use a T-table. The table has been started for you, below, with Alex's work already recorded in it. Add the numbers for the third, fourth, and fifth parts of the pattern.

Pattern Step	Number of Tiles Needed
1	1
2	4
3	

6. Generalize the pattern. How would you know how many tiles you need to build the 10th part of the pattern *without building it*?

Choose a variable, such as "*n*." Write an algebraic expression to determine the number of tiles you need for *any* part of the pattern.

_____

7. Be sure your name is written at the top of this handout before handing it in.

RESOURCES & LITERATURE CONNECTIONS

Sources

Classroom-Ready Materials Kits

Carolina Biological Supply® is the exclusive distributor of fully prepared GEMS Kits®, which contain all the materials you need for full classroom presentation of GEMS units. For more information, please visit www.carolina.com/GEMS or call (800) 227-1150.

Base Ten Blocks

The Math Learning Center
P.O. Box 3226
Salem, OR 97302
(800) 575-8130
www.mlc.pdx.edu

Educators Outlet
P.O. Box 397
Timnath, CO 80547
(800) 315-2212
www.educatorsoutlet.com

Related Curriculum Material

Family Math: The Middle School Years
Algebraic Reasoning and Number Sense
by Virginia Thompson and Karen Mayfield-Ingram
EQUALS/Lawrence Hall of Science, Berkeley, CA
(1998; 280 pp.)

Although meant for students in middle school, this book's activities are designed to develop and reinforce algebraic thinking. The relevant topics of inverses, identity elements, and the distributive property are explored in several of the math activities.

Family Math II
Achieving Success in Mathematics
by Grace Davila Coates and Virginia Thompson
EQUALS/Lawrence Hall of Science, Berkeley, CA
(2003; 200 pp.)

Presents exciting new mathematics materials to help families learn and enjoy math together. Topics include algebra, number sense, geometry, and probability and statistics. *Family Math II* provides ways to boost a child's success and confidence in mathematics. Also includes easy-to-follow instructions on how to organize Family Math classes.

Finite Differences
by Dale Seymour and Margaret Shedd
Dale Seymour Publications
Pearson Learning/Pearson Education, Parsippany, NJ
(1997; 120 pp.)

The term "finite differences" refers to an approach to generalizing problems involving sequences or tables. This book presents finite-difference problems that increase in difficulty, and describes the systematic technique of subtracting differences. Reproducible worksheets, answers, and glossary are provided.

Groundworks: Algebraic Thinking
by Carole Greenes and Carol Findell
Creative Publications
Wright Group/McGraw-Hill, Bothell, WA

The Groundworks series allows teachers to begin laying algebraic groundwork in first grade and continue through seventh. These reproducible books are organized around the six big ideas of algebra: representation, proportional reasoning, balance, variable, function, and inductive reasoning. The books in the Algebraic Thinking category are for students in grades 1–3. Each grade level has a 128-page teacher resource book and a 32-page student workbook. These books serve as a great resource to follow this GEMS unit and provide additional practice in algebraic problem solving.

Groundworks: Algebra Puzzles and Problems
by Carole Greenes and Carol Findell
Creative Publications
Wright Group/McGraw-Hill, Bothell, WA

Similar to the listing above, but for students in grades 4–7.

Wollygoggles and Other Creatures
Problems for Developing Thinking Skills
by Thomas C. O'Brien
ETA/Cuisenaire, Vernon Hills, IL
(1997; 64 pp.)

This book for grades 3–12 provides a fun-filled way to teach students how to look for patterns and solve problems systematically. Each section invites students to discover the rule for solving a set of problems. Pages are perforated and an answer key is provided.

Nonfiction for the Teacher

Algebra for Dummies
by Mary Jane Sterling
Hungry Minds/Wiley, New York, NY
(2001; 384 pp.)

This book covers everything from fractions to quadratic equations. Includes real-world examples and story problems. And of course, there are no "dummies" in mathematics, just people who think they are—because they weren't taught in creative and effective ways!

Algebraic Thinking, Grades K–12
Readings from NCTM's School-Based Journals and Other Publications
edited by Barbara Moses
National Council of Teachers of Mathematics, Reston, VA
(1999; 392 pp.)

Helps teachers understand the development of algebraic thinking and the types of activities at different grade levels that can foster such thinking in children. A

comprehensive collection of 59 specially selected articles from various NCTM publications and published works of other organizations. Defines algebraic thinking, analyzes the algebra curriculum, and supplies many classroom activities dealing with patterns, functions, and technology. Also discusses assessment, research issues, and professional development.

Navigating through Algebra in Grades 3–5
by Gilbert J. Cuevas and Karol Yeatts
National Council of Teachers of Mathematics, Reston, VA
(2001; 96 pp.)

The Navigations series is designed to provide ideas, activities, and materials to support the implementation of *Principles and Standards for School Mathematics.* Activities introduce and promote familiarity with important ideas of algebra, such as patterns, variables and equations, and functions. Features margin notes with teaching tips, anticipated student responses, assessment ideas, and references to some of the resources on the CD-ROM included with the book.

Principles and Standards for School Mathematics
National Council of Teachers of Mathematics, Reston, VA
(2000; 402 pp.)

An important tool and guide for teachers and education professionals. Incorporates a clear set of principles and an increased focus on how student knowledge grows as shown by recent research. Also includes ways to incorporate the use of technology to make mathematics instruction relevant and effective in a technological world.

Radical Equations
Math Literacy and Civil Rights
by Robert P. Moses and Charles E. Cobb, Jr.
Beacon Press, Boston, MA
(2001; 256 pp.)

Bob Moses's work to organize black voters in Mississippi famously transformed the political power of entire communities. Nearly 40 years later, Moses is organizing again, this time as teacher and founder of the national math literacy program called the Algebra Project. Moses suggests that the crisis in math literacy in poor communities is as urgent as the crisis of political access in

Mississippi in 1961. Through personal narrative and impassioned argument, he shows the lessons of the civil rights movement at work in a remarkable educational movement today. See also Internet sites section for more on the Algebra Project.

A Survey of Mathematics: Elementary Concepts and Their Historical Development
by Vivian Shaw Groza
(1968; out of print)

Although this book is out of print, it can be purchased online. An excellent resource book for teachers that's easy to read and understand. Contains useful historical and background information on mathematics.

Magazines

"Building Explicit and Recursive Forms of Patterns with the Function Game," *Mathematics Teaching in the Middle School,* April 2002, Vol. 7, Issue 8, p. 426.

"Children's Understanding of Equality: A Foundation for Algebra," *Teaching Children Mathematics,* December 1999, Vol. 5, No. 4, pp. 232–236.

"Developing Algebraic Reasoning in the Elementary Grades," *Teaching Children Mathematics,* December 1998, Vol. 4, No. 4, pp. 225–229.

"Experiences with Patterning," *Teaching Children Mathematics,* February 1997, Vol. 3, No. 6, pp. 282–288.
　　The entire issue focuses on algebraic thinking. This article is especially pertinent.

"The Function Game," *Mathematics Teaching in the Middle School,* November–December 1996, Vol. 2, No. 2, pp. 74–78.

Research Brief

"Algebra in the Elementary Grades," *in Brief,* Fall 2000, Vol. 1, No. 2.
　　in Brief is a publication of the National Center for Improving Student Learning and Achievement in Mathematics and Science, at the Wisconsin Center for Education Research, University of Wisconsin-Madison. It is available free upon request or it can be downloaded as a pdf file at www.wcer.wisc.edu/ncisla/publications/briefs/fall2000.pdf

Nonfiction for Students

Algebra to Go
A Mathematics Handbook
by the Great Source Education Group Staff
Great Source Education Group/Houghton Mifflin, Boston, MA
(2000; 523 pp.)

Presents key and often complex math topics in a way that's clear and easily understandable—from numeration and number theory to estimation, linear and non-linear equations, geometry, and data analysis. Provides detailed explanations, accessible charts and graphs, and examples to help students in grades 8 and above understand and retain algebraic concepts.

Fractals, Googols and Other Mathematical Tales
by Theoni Pappas
World Wide Publishing/Tetra, San Carlos, CA
(1993; 64 pp.)

Includes short stories and discussions which present such mathematical concepts as decimals, tangrams, number lines, and fractals. Offers an amusing and entertaining way to explore and share mathematical ideas—for those who love mathematics as well as those who hate it.

G is for Googol
A Math Alphabet Book
by David M. Schwartz;
illustrated by Marissa Moss
Tricycle Press, Berkeley, CA
(1998; 57 pp.)

Explains the meaning of mathematical terms which begin with the letters of the alphabet from abacus to a zillion. The meaning of "X" is particularly apt for this unit.

The Grapes of Math
Mind Stretching Math Riddles
by Greg Tang;
illustrated by Harry Briggs
Scholastic, New York, NY
(2001; 40 pp.)

This innovative book challenges children—and parents—to open their minds and solve problems in new and unexpected ways. By looking for patterns, symmetries, and familiar number combinations displayed within eye-catching pictures, math skills and young minds grow.

Math at Hand
A Mathematics Handbook
by the Great Source Education Group Staff
Great Source Education Group/Houghton Mifflin, Boston, MA
(1999; 548 pp.)

Provides a resource for students in grades 5–6 to find information on the math topics they're learning about—from mental math and problem solving to graphing, statistics, and probability to pre-algebra and geometry.

Math on Call
by the Great Source Education Group Staff
Great Source Education Group/Houghton Mifflin, Boston, MA
(1997; 608 pp.)

"Portable memory" for middle schoolers—a place to find information on math topics they're unsure of, look up a rule they may have forgotten, pick up study tips, even learn about computer spreadsheets and databases.

From introducing new skills to providing information for classroom projects to reinforcing concepts before an exam, this user-friendly handbook is packed with everything students in grades 6–8 need to know about math. Contains numerous examples with detailed explanations and easy-to-follow charts.

Math Talk
Mathematical Ideas in Poems for Two Voices
by Theoni Pappas
World Wide Publishing/Tetra, San Carlos, CA
(1991; 72 pp.)

Presents mathematical ideas through poetic dialogues intended to be read by two people. It may seem an odd combination, but as the foreward of the book states "Learning takes place via all our senses and by all forms of communication. Mathematical ideas can be learned through art, reading, conversations, lectures. Therefore, why not link mathematical ideas and poetic dialogues?" Particularly relevant are the poems about operations, variables, and the number 1.

Math to Know
A Mathematics Handbook
by Mary C. Cavanagh
Great Source Education Group/Houghton Mifflin, Boston, MA
(2000; 483 pp.)

This comprehensive resource provides clear explanations and numerous examples to help students in grades 3–4 understand important math concepts including basic operations, mental math and estimation, fractions and decimals, algebra and geometry, and graphing, statistics and probability. Contains an especially helpful chapter on algebraic thinking.

Powers of Ten
by Philip Morrison and Phylis Morrison
W.H. Freeman, New York, NY
(1982; 150 pp.)

Based on the well-known film *Powers of Ten* by the Office of Charles and Ray Eames. The book shows many different views of the same picnic—each one a power of ten different from the views it borders. The views extend from one billion light years away from

Earth, travel through our familiar daily scale, and go on to the microscopic. As such, it explores the concepts of scale and exponential growth. See also the Video and Internet sites sections.

Fiction for Students

A Gebra Named Al
by Wendy Isdell
Free Spirit Publishing, Minneapolis, MN
(1993; 128 pp.)
Grades 4–8

Trouble with her algebra homework leads Julie through a mysterious portal into the Land of Mathematics, where a zebra-like creature and horses representing Periodic Elements help her learn about math and chemistry in order to get home. The author wrote a draft of the book while in the eighth grade and published it when she was a senior in high school.

A Grain of Rice
by Helena C. Pittman
Bantam Books, New York, NY
(1992; 76pp.)
Grades 2–5

A clever, cheerful, hard-working farmer's son wins the hand of the emperor's daughter by outwitting the father who treasures her more than all the rice in China. Pong Lo's winning strategy is to use a mathematical ruse, asking simply for a grain of rice that is to be doubled every day for one hundred days. The book clearly illustrates exponential growth.

Amanda Bean's Amazing Dream
A Mathematical Story
by Cindy Neuschwander;
illustrated by Liza Woodruff;
math activities by Marilyn Burns
Scholastic Press, New York, NY
(1998; 40 pp.)
Grades K–3

Amanda loves to count everything, but not until she has an amazing dream does she finally realize that being able to multiply will help her count things faster. A Marilyn Burns Brainy Day Book.

Among the Odds & Evens
A Tale of Adventure
by Priscilla Turner;
illustrated by Whitney Turner
Farrar Straus Giroux, New York, NY
(1999; 32 pp.)
Grades K–3

When X and Y crash in the land of Wontoo, they cannot understand how the Numbers live the way they do, until they not only get used to it, but decide they want to stay in Wontoo. Introduces number concepts.

Anno's Mysterious Multiplying Jar
by Masaichiro and Mitsumasa Anno
Philomel/Putnam & Grosset, New York, NY
(1983; 44 pp.)
Grades 3–8

The simple text and illustrations introduce the mathematical concept of factorials. Through an understanding of multiplication, the reader can learn about factorials and the way that numbers can expand. On a second reading of the book, students can follow along using calculators to verify the large number of jars at the end of the story.

The King's Chessboard
by David Birch;
illustrated by Devis Grebu
Dial Books, New York, NY
(1988; 32 pp.)
Grades K–6

A proud king, too vain to admit what he does not know, learns a valuable (and exponential) lesson when he readily grants his wise man a special request. One grain of rice on the first square of a chessboard on the first day, two grains on the second square on the second day, four grains on the third square on the third day, and so on. After several days the counting of rice grains gives way to weighing, then the weighing gives way to counting sackfuls, then to wagonfuls. The king soon realizes that there is not enough rice in all of India to fulfill the wise man's request. This tale involves exponential growth and connects to the mathematical strands of number, pattern, and function. Students can use manipulatives in the classroom to see how quickly the rice amasses.

Math Curse

by Jon Scieszka;
illustrated by Lane Smith
Viking, New York, NY
(1995; 32 pp.)
Grades 2–Adult

When the teacher, Mrs. Fibonacci, tells her class that they can think of almost everything as a math problem, one student acquires a math anxiety that becomes a real curse. She thinks of every aspect of her home and school life as a number problem or riddle. After a long day overloaded with problems, she goes to bed and dreams the solution, which allows her to escape the math curse. Students can solve many of the problems posed in the book.

Melisande

by E. Nesbit;
illustrated by P.J. Lynch
Harcourt Brace Jovanovich, San Diego, CA
(1989; 42 pp.)
Grades 1–8

Princess Melisande will grow up to be bald because of a curse by an evil fairy. Upon being granted one wish, she asks for golden hair a yard long that will grow an inch every day and twice as fast when cut. Soon the princess realizes the implications of her wish. With the help of a determined godmother and a prince, order is restored. Though traditional fairy tale roles prevail, this story lends itself to an exploration of geometric progression (binomial sequence). Students can use yarn as a hands-on tool to understand how Melisande's hair grows.

Once Upon a Dime
A Math Adventure

by Nancy Kelly Allen;
illustrated by Adam Doyle
Charlesbridge Publishing, Watertown, MA
(1999; 32 pp.)
Grades 3–5

Organic farmer Truman Worth discovers that money grows on a special tree on his farm. The tree produces different kinds of money depending on what animal fertilizer he uses. Readers can practice math facts as they calculate the changing value of his crop.

One Riddle, One Answer

by Lauren Thompson;
illustrated by Linda S. Wingerter
Scholastic Press, New York, NY
(2001; 32 pp.)
Grades 3–6

A sultan's daughter, who loves numbers and riddles, creates a riddle using facts about numbers and operations to find the man who is best suited to be her husband. A full explanation of the solution is given at the end of the book, along with real-world uses for mathematics and a brief history of math concepts.

The Phantom Tollbooth

by Norton Juster;
illustrated by Jules Feiffer
Random House, New York, NY
(1989; 256 pp.)
Grades 2–8

Milo has mysterious and magical adventures when he drives his car past The Phantom Tollbooth and discovers The Lands Beyond. On his journey, Milo encounters amusing situations that involve numbers, geometry, measurement, and problem solving. The continuous play on words is delightful.

Sir Cumference and the Dragon of Pi
A Math Adventure

by Cindy Neuschwander;
illustrated by Wayne Geehan
Charlesbridge Publishing, Watertown, MA
(1999; 32 pp.)
Grades 1–4

When Sir Cumference drinks a potion that turns him into a dragon, his son Radius searches for the magic number known as pi, which will restore him to his former self. Helps students realize that pi is a ratio.

Spaghetti and Meatballs for All!
A Mathematical Story
by Marilyn Burns;
illustrated by Debbie Tilley
Scholastic, New York, NY
(1997; 40 pp.)
Grades 3–6

The seating for a family reunion gets complicated as people rearrange the tables and chairs to seat additional guests. A great book for presenting ideas about area and perimeter in a real-world context. Includes a discussion of the mathematics involved, and extensions for learning.

Videos

Powers of Ten
made by the office of Charles and Ray Eames for IBM;
produced by Eames Demetrois and Shelley Mills
Pyramid & Video, Santa Monica, CA
1989; 21 minutes

Dealing with scale, proportion, and dimension, the film moves in real time over a course of 40 powers of ten, from the cosmic distances of the universe to the heart of the atom. Demonstrates the relative size of things and what the addition of zero to any number means. Includes the original version of *Powers of Ten,* produced in 1968, entitled "A rough sketch for a proposed film dealing with the powers of ten and the relative size of things in the Universe." See also the Nonfiction for Students and Internet sites sections.

Stand and Deliver
a Menendez/Musca and Olmos production
Warner Home Video, Burbank, CA
1988; 103 minutes

Based on the true story of Jaime Escalante, a math teacher at Garfield High School in East Los Angeles, who refuses to write off his inner-city students as losers. Escalante pushes and inspires 18 students who were struggling with math to become math whizzes.

Internet sites

Note: While we do our best to provide long-lived addresses in this section, websites can be mercurial! Comparable alternative sites can generally be found with your Web browser.

The Algebra Project
www.algebra.org

Founded by civil rights activist and math educator Robert P. Moses in the 1980s, the Algebra Project is a national mathematics literacy effort aimed at helping low-income students and students of color. The project has developed curricular materials, trained teachers and trainers of teachers, and provided ongoing professional development support and community involvement activities to schools seeking to achieve a systemic change in mathematics education, particularly for African American and Latino/a students.

Ask Dr. Math
http://mathforum.org/dr.math/

This site is a question-and-answer service for math students and their teachers. The Frequently Asked Questions (FAQ) page is particularly useful. Also has a Teacher2Teacher section—a resource for teachers and parents who have questions about teaching mathematics.

Family Education Network
www.funbrain.com

Offers games for K–8 students as well as resources for teachers and parents. These two games are particularly useful.

Change Maker
www.funbrain.com/cashreg/index.html

A fun simulation in which students can practice making change for a dollar, five dollars, or 100 dollars. The amount of change is added to a piggy bank if the answer is correct, and subtracted if it's not.

MathCar Racing
www.funbrain.com/osa/index.html

Players use answers to math problems to advance their car around a racetrack. Involves mental math and strategy.

Math Forum Pre-Algebra Problem of the Week

http://mathforum.org/prealgpow/

The Math Forum's Problems of the Week (POWs) are designed to provide creative, non-routine challenges for students in grades 3–12. Problem-solving and mathematical communication are key elements of every problem. The pre-algebra POW is for students learning algebraic reasoning, identifying and applying patterns, ratio and proportion, and geometric ideas such as similarity.

The National Center for Improving Student Learning and Achievement in Mathematics and Science at the Wisconsin Center for Education Research, University of Wisconsin-Madison

www.wcer.wisc.edu/ncisla

This center is working with teachers and schools to study and develop ways to advance K–12 students' learning of mathematics and science. Their work is yielding new visions for student achievement and professional development programs that strengthen teachers' content knowledge and in-class practices. The website contains information on education research, publications, and teacher resources.

National Council of Teachers of Mathematics (NCTM)

www.nctm.org

NCTM is a professional organization for teachers of mathematics in grades K–12. The site has resources, lessons, and activities for teachers; information on conferences and other events; descriptions of their many publications; and information on professional development.

Powers of Ten

www.powersof10.com

Has information on the film and the book, and includes a poster and even an interactive CD-ROM.

Tower of Hanoi Puzzle

www.lhs.berkeley.edu/Java/Tower/Tower.html

On the Lawrence Hall of Science website, players look for a pattern in the sequence of moves to solve the puzzle. Please note that the puzzle represents a function with exponential growth that is beyond most students in grades 3–5; we include it here in case you find it useful for yourself or your advanced students.

REVIEWERS

We warmly thank the following educators, who reviewed, tested, or coordinated the trial tests for *Algebraic Reasoning, Electric Circuits, Invisible Universe,* and *Living with a Star* in manuscript or draft form. Their critical comments and recommendations, based on classroom presentation of these activities nationwide, contributed significantly to this GEMS publication. (The participation of these educators in the review process does not necessarily imply endorsement of the GEMS program or responsibility for statements or views expressed.) Classroom testing is a recognized and invaluable hallmark of GEMS curriculum development; feedback is carefully recorded and integrated as appropriate into the publications. WE THANK THEM ALL! ■

ARIZONA

Liberty School District #25, Buckeye
Wayne Bryan★
Terri Matteson

Arrowhead Elementary School, Phoenix
Noel Fasano
Jorjanne Miller
Kimberly Rimbey★
Delores Salisz
George Sowby
Coreen Weber

Hohokam Middle School, Tucson
Maria Federico-Brummer
Jennine Grogan★

Pistor Middle School, Tucson
Mike Ellis★

ARKANSAS

Carl Stuart Middle School, Conway
Chris Bing
Linda Dow
Gene Hodges
Charlcie Strange★

CALIFORNIA

Albany Middle School, Albany
Kay Sorg★

Rio Del Mar Elementary School, Aptos
Chris Ferrero
Doug Kyle
Tom LaHue★
Debbie Lawheed

Endeavor Elementary, Bakersfield
Matthew Diggle
Jan Karnowski★
Carolyn Reinen
Julie Rosales

Le Conte Elementary, Berkeley
Carole Chin★
Lorna Cross
Jennifer Smallwood

Longfellow Middle School, Berkeley
Karen Bush★

Juan Crespi Junior High School, El Sobrante
Randa Emera★
Juli Goldwyn
Geri Lommen
Julie Skow★

Oak Manor School, Fairfax
Celia Cuomo

Lorin Eden Elementary School, Hayward
Donna Goldenstein★
Elise Tran

M. H. Stanley Intermediate School, Lafayette
Tina Woodworth

Altamont Creek Elementary, Livermore
Pauline Huben★
Janice Louthan

Emma C. Smith Elementary, Livermore
J. Gulbransen

Leo R. Croche Elementary, Livermore
Corinne Agurkis

Mammoth Elementary School, Mammoth Lakes
Sue Barker★
Sandy Bramble
Stacey Posey
Janis Richardson

Cypress Elementary, Newbury Park
Cheryl Bowen
Christina Myren★
Kim Thompson